T0134815

METHODS IN MOLECULAR BIOLOGY

Series Editor
John M. Walker
School of Life and Medical Sciences
University of Hertfordshire
Hatfield, Hertfordshire, UK

For further volumes
http://www.springer.com/series/7651

Biomaterials for Tissue Engineering

Methods and Protocols

Edited by

Kanika Chawla

Cellerant Therapeutics, San Carlos, CA, USA

 Humana Press

Editor
Kanika Chawla
Cellerant Therapeutics
San Carlos, CA, USA

ISSN 1064-3745   ISSN 1940-6029   (electronic)
Methods in Molecular Biology
ISBN 978-1-4939-9271-3   ISBN 978-1-4939-7741-3   (eBook)
https://doi.org/10.1007/978-1-4939-7741-3

Dedication

To my father—who instilled in me my love of science, taught me to never stop asking questions, and inspires me still.

Preface

Biomaterials offer widespread opportunities for treating disease, and are being implemented in new tissue engineering and regenerative medicine-based therapies. The field of biomaterials had its inception in the late 1930s, with the development and application of poly(methyl methacrylate) (PMMA) in dentistry[1]. This was followed by joint and cardiovascular implants in the 1970s and 1980s and, with the advent of tissue engineering in the late 1980s, implantable organs by the early 2000s. The field itself is incredibly diverse with various materials (natural and synthetic) and cell types utilized and subjected to a wide set of stimuli (biochemical, biophysical), in order to generate an appropriate treatment for repair/replacement of biological tissues. Biomaterials are also emerging as extracellular-mimicking platforms designed to provide instructive cues to control cell behavior and, ultimately, be applied as solutions for biological problems.

Given an increasing aging population worldwide, an increase in the incidence of disease and procedures, as well as the development of new technologies, the global biomaterials market is estimated to be nearly US $150 billion by 2021. With this growth and need for applicable biomaterials for tissue engineering and regenerative medicine purposes, the design of future functional biomaterials will require collaboration with clinicians and physicians in order to determine the best and most relevant designs, blending both practical engineering and biological principles. A remaining challenge will be translation of the biomaterial (acellular or cell-laden, depending on the application) from the in vitro to the in vivo environment. Evaluation of the interaction of the biomaterial with the in vivo environment will be critical to successful implantation.

This volume of Methods in Molecular Biology—*Biomaterials for Tissue Engineering: Methods and Protocols*—contributes to that effort. In it, researchers provided step-by-step protocols for generation of various biomaterials for tissue engineering and regenerative medicine applications. The protocols described review a range of biomaterials including hydrogels and other matrices (natural, synthetic, self-healing), biomaterials for drug and gene delivery, surface modification and functionalization of biomaterials, and techniques for controlling biomaterial geometry, such as three-dimensional printing and electrospinning. The protocols also describe a variety of characterization techniques. Applications utilizing multiple cell types are also described. The target audiences of this book are scientists and engineers working in the areas of biomaterials, tissue engineering, and regenerative medicine. *Biomaterials for Tissue Engineering* is part of a larger, critically acclaimed Methods in Molecular Biology series. The entire series provides step-by-step protocols aimed at assisting the biological scientist or engineer in performing relevant studies in a reproducible manner.

---

[1] Ratner B, Hoffman A, Schoen F, Lemons J (2012) Biomaterials science: an introduction to materials in medicine, 3rd edn. Academic Press, 1573 pp.

The book comprises 15 chapters covering the aforementioned topics. I am pleased to provide a summary of author contributions below:

- Masehi-Lano and Chung describe a method for bone regeneration using citric acid-based scaffolds. The method is novel in that it utilizes low-pressure foaming to synthesize a scaffold with tailorable mechanical properties, similar to an elastomeric tissue.

- Shi et al. explore a versatile molecular design via self-assembling peptide-based micelles containing a cancer targeting sequence on their surface for targeted intracellular delivery, with potential application as a novel cancer therapeutic. Through rational design, these peptide amphiphiles can self-assemble into a variety of carrier nanostructures with controlled shape, size, and biological functionality to deliver therapeutic and imaging molecules to target cells.

- Chow describes an electrospinning methodology with functionalization strategy for generating scaffolds incorporating multiple bioactive cues for tissue engineering applications. The chapter provides a useful method for modifying poly(ε-caprolactone) (PCL) with peptides and electrospinning these peptide-PCL conjugates to functionalize a scaffold surface in a single step.

- Papastavrou et al. provide a unique low-temperature deposition modeling technique for preparing hierarchical ceramic scaffolds with tailorable porosity for applications in bone tissue engineering. The authors have developed a material with viscoelastic properties which can be tuned as well as a modified 3D printer to extrude the scaffolds at subzero temperatures.

- Zhou and Shikanov and Mendez examine two methodologies for studies involving ovarian follicles. The first focuses on a hydrogel-based method for studying the effects of toxicants on ovarian follicles while the second utilizes a polyethylene glycol (PEG) hydrogel-based system for engineering the ovarian follicle environment in order to study folliculogenesis.

- Chandrawati explores a versatile layer-by-layer assembly technique for generating polymer capsules as carriers for therapeutic intracellular delivery. Various materials can be incorporated in the multilayer design based on the properties and functionalities desired.

- Sproul et al. provide a methodology for modifying fibrin-based materials in order to control and modulate polymerization dynamics and promote cell infiltration, an aspect currently lacking in the application of fibrin tissue sealants. The chapter also describes methods for characterizing fibrin network morphology and polymerization and degradation dynamics.

- Paez et al. contribute a protocol for preparing thin, mechanically defined poly(acrylamide)-based hydrogels for use as model substrates in studying the cell-matrix interface. The methodology described can be used to incorporate a variety of instructive signals in order to more closely mimic in vivo scenarios.

- Gu et al. describe a method for fabricating neural tissue by three-dimensional printing neural stem cells with a bioink (comprised of alginate, chitosan, and agarose), and subsequent gelation of the bioink for cell encapsulation and eventual differentiation into functional neurons and neuroglia. The bioink strategy enables greater control over the three-dimensional architecture of the construct and supports high cell viability with efficient induction to neurons and other cells in the constructs.

- Vorwald et al. provide a high-throughput methodology for encapsulating mesenchymal stem cells (MSCs) within alginate spheroids in situ. The method results in improvement of retention and survival of cells upon transplantation, localizes payloads at the defect site, and provides cells with continued exposure to the properties of the biomaterial.

- Chao reports a robust method using electrospinning and post-processing to generate parallel polymer fibers with crimp to simulate the structure-function relationship of collagen fibers. The resulting platform can be used to study cell-material interactions in a biomimetic physical microenvironment mimicking native structure-function relationships.

- Witherel et al. contribute a protocol for culturing macrophages (a major component of the immune system and contributor in the in vivo response to implanted biomaterials) on three-dimensional biomaterials in order to study cell-material relationships which may influence the success of implantable biomaterials.

- Paschuk provides a relevant methodology for generating self-assembled peptide-based hydrogels by solid phase peptide synthesis, including suggestions on how to improve synthetic yield and purity of self-assembling peptides.

- Kaur et al. describe a peptide amphiphile hydrogel designed to release hydrogen sulfide, an endogenously produced cell signaling molecule that has a role in several physiological processes including angiogenesis and inflammation, in a controlled manner.

I would like to acknowledge the authors for contributing their time and energy to this edition. I wish to acknowledge John Walker for his guidance in editing the book chapters. I am also very thankful for my husband Nimeesh's encouragement; this book would not have been possible without his support. Our sons, Deven and Arjun, represent significant inspiration in working to develop novel biomaterials for tissue engineering and regenerative medicine applications—in the hopes of benefiting the next generation. I would also like to express my gratitude to my parents, brother, and family for their encouragement.

San Carlos, CA, USA   *Kanika Chawla*

Contents

Contributors

HELENA S. AZEVEDO · *School of Engineering and Materials Science, Institute of Bioengineering, Queen Mary University of London, London, UK*

P. BREEDON · *School of Science and Technology, Nottingham Trent University, Nottingham, England, UK*

ASHLEY C. BROWN · *Joint Department of Biomedical Engineering, North Carolina State University and The University of North Carolina at Chapel Hill, Raleigh, NC, USA*

ARÁNZAZU DEL CAMPO · *INM—Leibniz Institute for New Materials, Saarbrücken, Germany; Chemistry Department, Saarland University, Saarland, Germany*

RONA CHANDRAWATI · *School of Chemical Engineering, The University of New South Wales, Sydney, NSW, Australia*

PEN-HSIU GRACE CHAO · *Institute of Biomedical Engineering, National Taiwan University, Taipei, Taiwan*

LESLEY W. CHOW · *Department of Materials Science and Engineering, Lehigh University, Bethlehem, PA, USA; Department of Bioengineering, Lehigh University, Bethlehem, PA, USA*

EUN JI CHUNG · *Department of Biomedical Engineering, University of Southern California, Los Angeles, CA, USA*

JEREMY M. CROOK · *ARC Centre of Excellence for Electromaterials Science, Intelligent Polymer Research Institute, AIIM Facility, University of Wollongong, Wollongong, NSW, Australia; Illawarra Health and Medical Research Institute, University of Wollongong, Wollongong, NSW, Australia; Department of Surgery, St Vincent's Hospital, The University of Melbourne, Fitzroy, VIC, Australia*

HONGGANG CUI · *Department of Chemical and Biomolecular Engineering, Institute for NanoBioTechnology, Johns Hopkins University, Baltimore, MD, USA*

D. FAIRHURST · *School of Science and Technology, Nottingham Trent University, Nottingham, England, UK*

ALEEZA FARRUKH · *INM—Leibniz Institute for New Materials, Saarbrücken, Germany*

PAMELA L. GRANEY · *Biomaterials & Regenerative Medicine Laboratory, School of Biomedical Engineering, Science, and Health Systems, Drexel University, Philadelphia, PA, USA*

QI GU · *ARC Centre of Excellence for Electromaterials Science, Intelligent Polymer Research Institute, AIIM Facility, University of Wollongong, Wollongong, NSW, Australia*

RILEY T. HANNAN · *Department of Pathology, University of Virginia, Charlottesville, VA, USA*

STEVE S. HO · *Department of Biomedical Engineering, University of California—Davis, Davis, CA, USA*

KULJEET KAUR · *Department of Chemistry, Virginia Tech University, Blacksburg, VA, USA*

J. KENT LEACH · *Department of Biomedical Engineering, University of California—Davis, Davis, CA, USA; Department of Orthopaedic Surgery, School of Medicine, University of California—Davis, Sacramento, CA, USA*

RAN LIN · *Department of Chemical and Biomolecular Engineering, Institute for NanoBioTechnology, Johns Hopkins University, Baltimore, MD, USA*

JACQUELINE J. MASEHI-LANO · *Department of Biomedical Engineering, University of Southern California, Los Angeles, CA, USA*

JOHN B. MATSON · *Department of Chemistry, Virginia Tech University, Blacksburg, VA, USA*

UZIEL MENDEZ · *Department of Biomedical Engineering, University of Michigan, Ann Arbor, MI, USA*

JULIETA I. PAEZ · *INM—Leibniz Institute for New Materials, Saarbrücken, Germany*

E. PAPASTAVROU · *School of Science and Technology, Nottingham Trent University, Nottingham, England, UK*

EVA TOMASKOVIC-CROOK · *ARC Centre of Excellence for Electromaterials Science, Intelligent Polymer Research Institute, AIIM Facility, University of Wollongong, Wollongong, NSW, Australia; Illawarra Health and Medical Research Institute, University of Wollongong, Wollongong, NSW, Australia*

E. THOMAS PASHUCK · *Department of Materials Science and Engineering, Imperial College London, London, UK*

YUN QIAN · *Department of Chemistry, Virginia Tech University, Blacksburg, VA, USA*

YEJIAO SHI · *School of Engineering and Materials Science, Institute of Bioengineering, Queen Mary University of London, London, UK*

ARIELLA SHIKANOV · *Department of Biomedical Engineering, University of Michigan, Ann Arbor, MI, USA; Department of Macromolecular Science & Engineering, University of Michigan, Ann Arbor, MI, USA*

KARA L. SPILLER · *Biomaterials & Regenerative Medicine Laboratory, School of Biomedical Engineering, Science, and Health Systems, Drexel University, Philadelphia, PA, USA*

ERIN P. SPROUL · *Joint Department of Biomedical Engineering, North Carolina State University and The University of North Carolina at Chapel Hill, Raleigh, NC, USA*

EVA TOMASKOVIC-CROOK · *ARC Centre of Excellence for Electromaterials Science, Intelligent Polymer Research Institute, AIIM Facility, University of Wollongong, Wollongong, NSW, Australia; Illawarra Health and Medical Research Institute, University of Wollongong, Wollongong, NSW, Australia*

OYA USTAHÜSEYIN · *INM—Leibniz Institute for New Materials, Saarbrücken, Germany*

CHARLOTTE E. VORWALD · *Department of Biomedical Engineering, University of California—Davis, Davis, CA, USA*

GORDON G. WALLACE · *ARC Centre of Excellence for Electromaterials Science, Intelligent Polymer Research Institute, AIIM Facility, University of Wollongong, Wollongong, NSW, Australia*

JACKLYN WHITEHEAD · *Department of Biomedical Engineering, University of California—Davis, Davis, CA, USA*

CLAIRE E. WITHEREL · *Biomaterials & Regenerative Medicine Laboratory, School of Biomedical Engineering, Science, and Health Systems, Drexel University, Philadelphia, PA, USA*

HONG ZHOU · *Department of Biomedical Engineering, University of Michigan, Ann Arbor, MI, USA*

Chapter 1

Engineering Citric Acid-Based Porous Scaffolds for Bone Regeneration

Jacqueline J. Masehi-Lano and Eun Ji Chung

Abstract

Tissue engineering aims to develop scaffolds that are biocompatible and mimic the mechanical and biological properties of the target tissue as closely as possible. Here, we describe the fabrication and characterization of a biodegradable, elastomeric porous scaffold: poly(1,8-octanediol-*co*-citric acid) (POC) incorporated with nanoscale hydroxyapatite (HA). While this chapter focuses on the scaffold's potential for bone regeneration, POC can also be used in other tissue engineering applications requiring an elastomeric implant. Because of the relative ease with which POC can be synthesized, its mechanical properties can be tailored to mimic the structure and function of the target elastomeric tissue for enhanced tissue regeneration.

Key words Hydroxyapatite, Tissue engineering, Tissue regeneration, Osteogenicity, Composites, Mechanical properties

1   Introduction

The objective of tissue engineering is to develop biological scaffolds that restore, maintain, or improve the function of damaged tissue. Generally, damaged tissue is replaced by autografts (tissue derived from the patient) or allografts (tissue taken from a donor and implanted in a patient) [1, 2]. However, both techniques have major drawbacks such as donor-site morbidity in the case of autografts, and the possibility of disease transmission in the case of allografts. The advent of tissue engineering shifted the focus from tissue replacement to tissue regeneration by developing biological substitutes that act as templates to guide the formation of new tissue produced by the own body. More specifically, tissue substitutes are synthesized in vitro, implanted at the injured site, and gradually replaced with new native tissue by the body's own regenerative capacity in vivo [1].

Despite the recognized abundance of tissues of the body that have elastomeric properties such as bone, research into the

Kanika Chawla (ed.), *Biomaterials for Tissue Engineering: Methods and Protocols*, Methods in Molecular Biology, vol. 1758, https://doi.org/10.1007/978-1-4939-7741-3_1, © Springer Science+Business Media, LLC, part of Springer Nature 2018

development of compliant biodegradable scaffolds has been highly underserved and 90% of bone grafting procedures today still require autografts or allografts [2]. This is because few biocompatible synthetic materials meet the requirements of bone grafts and the materials reported in literature are too costly for clinical and commercial implementation [3]. For example, while existing biodegradable scaffolds such as poly(L-lactide) (PLL), poly(caprolactone) (PCL), or combinations of these polymers show favorable in vivo compatibility with progenitor cells and osteoblasts, PCL and PLL show mechanical properties that are either too weak for bone integration or have degradation rates that hamper tissue integration. Here, we describe the synthesis of a porous biodegradable elastomer, poly(1,8-octanediol-*co*-citric acid) (POC) incorporated with nanoscale hydroxyapatite (HA). POC offers many advantages compared to the other polymeric biodegradable scaffolds, namely, the use of nontoxic monomers, relatively easy synthesis that can be carried out under mild conditions, controllable mechanical and degradation properties, cost efficiency, and compatibility with several cell types [3].

Bone is a natural composite that is made up of HA crystals embedded within an elastic collagen matrix. HA makes up 60–70% of native bone by weight and incorporation of HA confers the scaffold with osteogenicity and osteoconductivity [4–7]. POC is an ideal polymer for the matrix phase, as it has previously been shown to accommodate up to 60–70% of HA particles by weight [3, 8]. In contrast to other polymer-HA composites, the elastomeric and binding properties of POC enable the HA component to make up the majority of the composite [4]. This, in turn, maximizes the osteoconductivity and more closely mimics the native components of bone [3, 8].

Incorporating porosity to the POC-HA scaffold further mimics the native structure of bone, which is 50–90% porous [4, 9–12]. Current methods to achieve porosity require the use of organic solvents, supercritical carbon dioxide, or porogen leaching, which are not applicable to many polymers. In contrast, interconnected pores in the POC-HA scaffolds can be fabricated using low-pressure foaming (LPF). In LPF, the air bubbles that arise during the polymer mixing step act as pore nucleation sites [13]. Applying vacuum expands the nucleation sites into interconnected pores that are stabilized through cross-linking.

Citric acid-based nanocomposites in both short- (6 weeks) and long-term (26 weeks) studies have previously shown favorable bone-implant and bone-cartilage response in rabbits [14, 15]. Furthermore, the degradation rate of porous POC-HA enables optimal tissue ingrowth and regeneration. In addition to the fabrication and characterization of porous POC-HA scaffolds, this chapter describes methods to observe human mesenchymal stem cell (hMSC) attachment and alkaline phosphatase activity to confirm scaffold biocompatibility and osteogenicity.

2 Materials

2.1 POC-HA

1. Citric acid (Sigma-Aldrich, St. Louis, MO, USA).
2. 1,8-Octanediol (Sigma-Aldrich, St. Louis, MO, USA).
3. 250 mL Round-bottom flask (Synthware Glass, Pleasant Prairie, WI, USA).
4. Magnetic stir bar (VWR, Radnor, PA, USA).
5. Hot plate with magnetic stirrer (VWR, Radnor, PA, USA).
6. Silicon oil bath (Sigma-Aldrich, St. Louis, MO, USA).
7. Paper towels.
8. Ethanol (200 Proof) (Sigma-Aldrich, St. Louis, MO, USA).
9. Parafilm M laboratory film (Bemis, Neenah, WI, USA).
10. Orbital shaker (VWR, Radnor, PA, USA).
11. Large beaker (VWR, Radnor, PA, USA).
12. Pasteur pipette (VWR, Radnor, PA, USA).
13. Rubber bulb for pipette (VWR, Radnor, PA, USA).
14. Lyophilizer (Labconco, Fort Scott, KS, USA).
15. Spatula (VWR, Radnor, PA, USA).
16. Amber glass bottle.

2.2 Viscosity

1. Ubbelohde viscometer (Cannon Instrument Co., State College, PA, USA).

2.3 HA

1. Hydroxyapatite nanoparticles (Berkeley Advanced Biomaterials, Inc., Berkeley, CA, USA).
2. Ethanol (200 Proof) (Sigma-Aldrich, St. Louis, MO, USA).
3. X-ray diffractometer.
4. Scanning electron microscope (SEM).
5. Teflon dishes (100 mL) (VWR, Radnor, PA, USA).

2.4 Compression Testing

1. Sintech mechanical tester model 20/G (Research Triangle Park, NC, USA).
2. Milli-Q water.

2.5 Porosity Measurements Method 1

1. Water bath (Branson, North Olmsted, OH, USA).
2. Milli-Q water.
3. Density bottle.
4. Thin wire.
5. Weigh scale.
6. Ethanol.

7. Kimwipes (VWR, Radnor, PA, USA).

8. Vacuum hose.

2.6 Porosity Measurements Method 2

1. Autopore IV (Micromeritics, Norcross, GA, USA).

2.7 Preparation of POC-Based Scaffolds for Human Mesenchymal Stem Cell (hMSC) Culture

1. hMSCs (Lonza, Walkersville, MD, USA).

2. Expansion media: Prepare low-glucose DMEM (1 g/mL glucose) with 10% fetal bovine serum (FBS) and 1% penicillin/streptomycin.

3. 24-Well plate.

4. AN74j/Anprolene ethylene oxide sterilization system (Anderson Sterilization, Inc., Haw River, NC, USA).

5. 2.5% Glutaraldehyde.

6. Ethanol.

7. Lyophilizer.

8. Sputter coater.

9. SEM.

10. Quant-iT Pico Green dsDNA Reagent (Invitrogen, Carlsbad, CA, USA).

11. 0.1% Triton-X 100.

12. Sonicator.

13. 10 mM p-Nitrophenyl phosphate.

14. 50 mM Glycine buffer.

15. 0.5 mM $MgCl_2$ (Sigma Aldrich, St. Louis, MO, USA).

16. Spectrophotometer.

3 Methods

3.1 Poly(1,8-octanediol-co-citric acid) Synthesis

1. Add equimolar amounts of citric acid (0.5 mol, 96.06 g) and 1,8-octanediol (0.5 mol, 73.11 g) to a 250 mL round-bottom flask. Put a magnetic stir bar inside the flask and place the flask in a silicon oil bath on a magnetic stirrer with a hot plate. Stir at 155 °C and 140 rpm for approximately 9.5 min to melt the mixture. Lower the temperature to 140 °C and stir the mixture at 140 rpm for 45 min to create the prepolymer solution. Continue to polymerize until the magnetic stir bar can no longer spin even at low speeds (*see* **Note 1**). While the polymerization process is taking place, wash a large beaker with ethanol and place a magnetic stir bar inside it. Weigh the beaker and

magnetic stir bar and record the empty weight. This will be used in the purification steps later.

2. Remove the flask from the oil bath. Wipe off excess oil from the flask with paper towels. Immediately place the flask under cold running water for 5–10 min until the flask cools to room temperature. Next, add ethanol to the flask such that the ratio of ethanol to POC is 3:1. Cover the flask with parafilm. Place the flask on an orbital shaker to dissolve the POC polymer (*see* **Note 2**).

3. Fill the previously weighed beaker containing the magnetic stir bar with water. Place the beaker on a magnetic stirrer. After the POC has dissolved, use a Pasteur pipette with a bulb to add the POC dropwise to the large beaker of water under magnetic stirring. The ratio of water to POC in ethanol should be approximately 2:1 so that any unreacted monomers and low-molecular-weight species can be dissolved in the water. Once all the POC prepolymer has been added to the water, stop mixing and allow the POC prepolymer to settle to the bottom of the beaker. Discard the water without letting any of the POC prepolymer pour out from the bottom of the beaker (*see* **Note 3**). After discarding the water for the final time, place the beaker with POC prepolymer in a −80 °C freezer. After freezing, place the beaker in a lyophilizer for 2–4 days to remove all the water (*see* **Note 4**).

4. Following the freeze-drying process, weigh the beaker containing the magnetic stir bar and POC to obtain the final weight. Subtract the empty weight from the final weight to the get the mass of POC recovered. Determine the recovery using the below equation:

$$\text{Recovery} = \frac{\text{Mass of POC recovered}}{\text{Mass of POC originally polymerized}} \times 100$$

5. Determine the mass of ethanol that must be added to the beaker to make a 30% (w/w) solution of POC in ethanol using the following equation:

$$0.30 = \frac{\text{Mass of POC recovered}}{\text{Mass of POC recovered} + \text{Mass of ethanol}}$$

Convert the mass of ethanol calculated from the above equation to a volume by using the density of ethanol (ρ_{ethanol} ~0.789 g/cm³ at room temperature). Add the correct amount of ethanol to the beaker. Cover the beaker with parafilm and place on an orbital shaker to dissolve the POC (*see* **Note 5**). After dissolving, pour the contents of the beaker into an amber glass bottle and store at room temperature until further use.

3.2 Fabrication of Porous POC-HA Nanocomposites via Low-Pressure Foaming

1. Dissolve 10 g of POC prepolymer in 18 g ethanol at 80 °C. Mix with the wt.% of HA powder. Warm Teflon dishes to 80 °C. Place the POC-HA mixture in the dishes. Stir the mixture manually with a spatula for 30 s every 5 min for 2 h. Mix manually for 30 s every hour for 24 h. During POC-HA mixing, the sample can be left covered overnight at room temperature and finished mixing the next day. This is only after the POC-HA mixture is more solid vs. liquid mixture. After the POC-HA becomes a solid that can be rolled into a ball, place the mass at 80 °C under vacuum at 2 Pa for 3 days. Then place the mass at 120 °C under vacuum at 2 Pa for 1 day (*see* **Note 6**). To obtain nonporous POC-HA composites, insert the POC-HA mass into Teflon molds to obtain the desired shape and polymerize at 80 °C for 2 Pa for 3 days without vacuum (Fig. 1).

3.2.1 Tuning of Prepolymer Viscosity on POC-HA Scaffolds

1. To fabricate pre-POC with varying viscosities and degrees of polymerization, synthesize the pre-POC under one of the following conditions: 9.5 min at 155 °C, 15 min at 155 °C, or 15 min at 155 °C followed by 35 min at 140 °C. Stir all three types of pre-POC at 140 rpm during synthesis.

2. Measure the kinematic viscosities at 25 °C with 2% pre-POC solution in ethanol (w/v) using an Ubbelohde viscometer, following the manufacturer's instructions. Stir the POC mixture manually as specified above and post-polymerize under the same conditions.

3.3 Compression Testing

Prepare at least three samples for each POC-HA scaffold. To prepare samples under wet testing, soak some of the samples in Milli-Q water for 24 h at room temperature before testing. Use a mechanical tester (i.e., Sintech mechanical tester model 20/G) to measure compression moduli of samples under dry and wet conditions according to American Society for Testing and Materials (ASTM) d695.

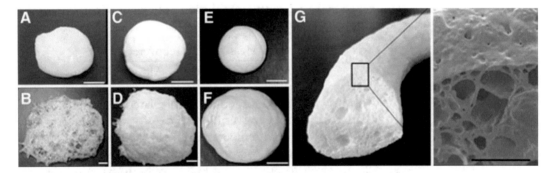

Fig. 1 Digital images of (**a**) solid POC, (**b**) porous POC, (**c**) solid 20% by weight HA nanocrystals in POC (POC-20HA), (**d**) porous POC-20HA, (**e**) solid POC-40HA, and (**f**) porous POC-40HA. (**g**) The porous scaffolds contain an outer nonporous layer (reproduced from [13] with permission from Mary Ann Liebert, Inc.)

3.3.1 Porosity Measurement of POC-HA Scaffolds Using Archimedes Principle (Method 1)

1. Fill the water bath with water. The level of water in the bath should cover the neck of the density bottle but not high enough that water will enter the bottle. Warm the bath to 30 °C. Place the thin wire in the plug of the cap of the density bottle to enable hanging of polymer samples from the end of the wire.

2. Cut out polymer samples and weigh them on a scale. Record the mass of each scaffold (W_s). Ensure that the polymer sample is small enough to fit through the neck of the density bottle but large enough to be easily placed on the hook of the density bottle. Fill the density bottle with ethanol. Place it in the water bath at 30 °C.

3. Equilibrate the density bottle containing ethanol in the water bath. When the bottle is initially placed in the water bath, the ethanol will expand as the temperature rises. This expansion will push ethanol out the hole in the top of the cap and form a droplet on the surface of the cap. When ethanol stops rising out the top of the cap and the top becomes dry, the bottle has reached equilibrium. After equilibrating to 30 °C, remove the density bottle from the water bath. Dry the outside with a Kimwipe. Place the density bottle on scale to obtain its mass filled with ethanol (W_1).

4. Place the scaffold on the hook of the density bottle such that it is hanging above the ethanol in the bottle. Attach a vacuum hose to the bottle cap and through repeated vacuum cycles, ensure that ethanol is filling up all the pores. The pores are filled when air bubbles no longer escape from the scaffold under vacuum.

5. Refill the density bottle with ethanol until it is full. Place it back in the water bath at 30 °C. When the bottle equilibrates again, immediately remove it from the water bath. Dry the bottle with a Kimwipe and place it on the scale to obtain the mass of the density bottle with the scaffold (W_2). Remove the scaffold from the bottle and weigh the bottle one more time to obtain the mass of the bottle after removing the scaffold (W_3).

The porosity of the scaffold can be calculated using the following equations (*see* **Note 7**):

$$V_P = \frac{(W_2 - W_3 - W_S)}{\rho_e}$$

$$V_S = \frac{(W_1 - W_2 + W_S)}{\rho_e}$$

$$\rho_S = \frac{W_S}{V_S}$$

$$\varepsilon = \frac{V_P}{V_P + V_S}$$

- V_p = Volume of scaffold pores
- V_s = Volume of scaffold skeleton
- ρ_s = Scaffold density
- ε = Porosity
- ρ_e = Density of ethanol

3.3.2 Porosity Measurement of POC-HA Scaffolds Using Mercury Intrusion Porosimetry and SEM (Method 2)

1. Use an Autopore IV to determine the incremental intrusion of mercury with pressure for POC-HA samples. Calculate a volume percentage for a particular pore diameter using the following equation:

$$\text{Volume percentage} = \frac{\text{Incremental intrusion} \times \text{Porosity}}{\text{Total intrusion}} \times 100$$

2. Sputter-coat the cross section and the surface of the scaffolds and observe using SEM. Measure the average pore size and pore wall thickness manually using the SEM images. Perform at least four measurements for each parameter (Fig. 2).

3.4 hMSC Culture on POC-HA and Osteogenesis

1. Culture and expand hMSCs in expansion media at 37 °C in humidified air containing 5% CO_2 (*see* **Note 8**).

2. Place each sample in a 24-well plate. Sterilize the samples using an AN74j/Anprolene ethylene oxide sterilization system that performs a 2-h degassing step under vacuum after 12 h of gas exposure. Prepare the samples as 10 mm cubes. Condition the samples in supplemented media at 37 °C for 24 h (*see* **Note 9**). Next, suspend 40,000 cells in 15 μL of media before seeding. 2 h post-seeding, aliquot 2 mL of media into each well.

3.4.1 Morphology and Attachment of hMSCs on POC-HA Scaffolds

1. After 24 h, fix the samples with 2.5% glutaraldehyde. Then dehydrate the samples in graded series of ethanol (i.e., 50, 60, 70, 80, 90, 95% ethanol) and freeze-dry them. Sputter-coat the samples and observe the morphology of the cells using SEM.

2. For hMSC attachment, quantify total DNA using Quant-iT Pico Green dsDNA Reagent. Lyse cells using 0.1% Triton-X 100. Sonicate for 20 min and use the lysate. Use at least three samples for each material.

3.4.2 Intracellular Alkaline Phosphatase Activity of hMSCs on POC-HA Scaffolds

Add one aliquot of the cell lysate to an equal amount of reaction buffer containing 10 mM p-nitrophenyl phosphate in 50 mM glycine buffer at pH 10.5, supplemented with 0.5 mM $MgCl_2$. After 30 min at 37 °C, add 0.05 M NaOH to stop the reaction. Measure the reaction solution by spectrophotometry at 410 nm. Construct a standard curve from different dilutions of p-nitrophenol stock solution. Use at least three samples for each scaffold type for each media condition and normalize results by total DNA.

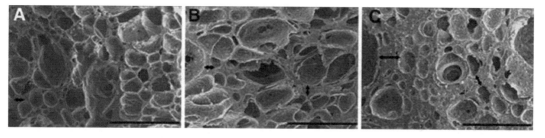

Fig. 2 SEM cross-section images of POC-40HA prepolymer synthesized under (**a**) 9.5 min at 155 °C, (**b**) 15 min at 155 °C, and (**c**) 15 min at 155 °C followed by 35 min at 140 °C. The distance between pores increased with increased polymerization time. Scale bar: 1 mm (reproduced from [13] with permission from Mary Ann Liebert, Inc.)

4 Notes

1. Stop at 100 rpm, when the stir bar can no longer spin, but moves very slowly with brief jolts. For 0.1 moles/monomer, the polymerization time should be between 1 and 1.5 h.

2. If the POC does not dissolve after 2 days and instead swells, the POC was over-cross-linked on the hot plate and the reaction was not stopped in time. Reduce the polymerization time on the hot plate next time and remove the hot plate before the stir bar can no longer spin.

3. POC can be redissolved in ethanol depending on the desired purity of the polymer.

4. The time in the lyophilizer will vary depending on the amount of POC prepolymer synthesized.

5. Some of the POC may stick to the sides of the beaker above the level of the ethanol. If the beaker is sealed tightly with parafilm then the POC on the sides will eventually dissolve from the ethanol vapor.

6. Perform SEM or transmission electron microscopy (TEM) to confirm the size of the HA microcrystals.

7. This method can be highly variable. Slight variations in the temperature, volume of ethanol in the density bottle when full, or whether or not the pores are completely filled with ethanol can cause large variations in the calculated porosity. This method will need to be repeated many times so that the technique can be mastered and the variability reduced. It is advisable to have an experience member of the lab guide you through this protocol the first time. In addition, if more accurate data is needed, mercury intrusion porosimetry (Subheading 3.3.2) should be implemented for porosity measurements.

8. Cells at passage five were used in the studies and media was changed every 3 or 4 days.

9. Make sure that media color does not change due to monomeric leaching.

Acknowledgments

This work was supported by NIH grant R00HL124279 granted to EJC.

References

1. O'Brien FJ (2011) Biomaterials and scaffolds for tissue engineering. Materials Today 14:88–95

2. Chung EJ, Sugimoto M, Ameer GA (2011) The role of hydroxyapatite in citric acid-based nanocomposites: surface characteristics, degradation, and osteogenicity in vitro. Acta Biomater 7:4057–4063

3. Yang J, Webb AR, Ameer GA (2004) Novel citric acid-based biodegradable elastomers for tissue engineering. Adv Mater 16:511–516

4. Chung EJ (2011) Poly(diol citrate)-hydroxyapatite nanocomposites for bone and ligament tissue engineering. Dissertation Northwestern University

5. Murugan R, Ramakrishna S (2005) Development of nanocomposites for bone grafting. Comp Sci Tech 65:2385–2406

6. Ignatius AA, Betz O, Augat P, Claes LE (2001) In vivo investigations on composites made of resorbable ceramics and poly(lactide) used as bone graft substitutes. J Biomed Mater Res 58:701–709

7. Rizzi SC, Heath DJ, Coombes AG, Bock N, Textor M, Downes S (2001) Biodegradable polymer/hydroxyapatite composites: surface analysis and initial attachment of human osteoblasts. J Biomed Mater Res 55(4):475–486

8. Qiu H, Yang J, Kodali P, Ameer GA (2006) A citric acid-based hydroxyapatite composite for orthopedic implants. Biomaterials 27:5845–5854

9. Móczó J, Pukánszky B (2008) Polymer micro and nanocomposites: structure, interactions, properties. J Indus Eng Chem 14:535–563

10. Sikavitsas VI, Temenoff JS, Mikos AG (2001) Biomaterials and bone mechanotransduction. Biomaterials 22:2581–2593

11. Temenoff JS, Lu L, Mikos AG (1999) Bone engineering. Em Squared, Toronto

12. Keaveny TM, Morgan EF, Niebur GL, Yeh OC (2001) Biomechanics of trabecular bone. Annu Rev Biomed Eng 3:307–333

13. Chung EJ, Sugimoto M, Koh JL, Ameer GA (2012) Low-pressure foaming: a novel method for the fabrication of porous scaffolds for tissue engineering. Tissue Eng Part C Methods 18:113–121

14. Chung EJ, Kodali P, Laskin W, Koh JL, Ameer GA (2011) Long-term *in vivo* response to citric acid-based nanocomposites for orthopaedic tissue engineering. J Mater Sci Mater Med 22:2131–2138

15. Chung EJ, Qiu H, Kodali P, Yang S, Sprague SM, Hwong J, Koh J, Ameer GA (2011) Early tissue response to citric acid-based micro- and nanocomposites. J Biomed Mater Res Part A 96:29–37

Multifunctional Self-Assembling Peptide-Based Nanostructures for Targeted Intracellular Delivery: Design, Physicochemical Characterization, and Biological Assessment

Yejiao Shi, Ran Lin, Honggang Cui, and Helena S. Azevedo

Abstract

Peptide amphiphiles (PAs), consisting of a hydrophobic alkyl chain covalently bound to a hydrophilic peptide sequence, possess a versatile molecular design due to their combined self-assembling features of amphiphile surfactants and biological functionalities of peptides. Through rational design, PAs can self-assemble into a variety of nanostructures with controlled shape, size, and biological functionality to deliver therapeutic and imaging agents to target cells. Here, we describe principles to design multifunctional PAs for self-assembly into micellar nanostructures and targeted intracellular delivery. The PA micelles are designed to display a tumour targeting sequence on their surfaces and direct their interactions with specific cells. This targeting sequence includes an enzymatic sensitive sequence that can be cleaved upon exposure to matrix metalloproteinase 2 (MMP-2), an enzyme overexpressed in tumor environment, allowing the presentation of a cell-penetrating domain. The presentation of this domain will then facilitate the delivery of therapeutics for cancer treatment inside targeted cells. Methods to characterize the key physicochemical properties of PAs and their assemblies, including secondary structure, critical micelle concentration, shape and size, are described in detail. The enzyme responsiveness of PA assemblies is described with respect to their degradation by MMP-2. Protocols to evaluate the cytotoxicity and cellular uptake of the micellar carriers are also included.

Key words Self-assembly, Multifunctional micelle, Enzyme-responsive, Cell-penetrating peptide, Targeted intracellular delivery

1 Introduction

Current research in the area of drug delivery is focused on the development of stimuli-sensitive systems that can perform multiple functions and overcome diverse physiological barriers to optimize delivery to target sites (organs, tissues, cells) [1]. The ubiquitous distribution and great diversity of enzymes present in the human body have motivated the design of nanocarriers responsive to enzyme activities and thus translate the enzymatic modification

Kanika Chawla (ed.), *Biomaterials for Tissue Engineering: Methods and Protocols*, Methods in Molecular Biology, vol. 1758, https://doi.org/10.1007/978-1-4939-7741-3_2, © Springer Science+Business Media, LLC, part of Springer Nature 2018

into a suitable response, such as material degradation or disassembly, and the simultaneous release of entrapped drug. Targeted delivery strategies exploit the selective affinity of ligands for tissue components or cell-surface receptors for retention in the targeted tissue and/or enhanced uptake by specific cells. It can reduce toxicity and adverse side effects by accumulating the delivered therapeutics specifically at the site of interest.

Molecular self-assembly provides the possibility to construct multifunctional nano-architectures with precise shape, size and functional control. It is a bottom-up approach which uses rationally designed molecular units to construct various types of nanostructures (micelles, vesicles, and fibers) by self-organization processes driven by non-covalent forces (electrostatic interactions, hydrogen bondings, aromatic interactions, van der Waals forces, and hydrophobic effects) [2]. Due to the involvement of non-covalent interactions, self-assembly is also a reversible process, and the self-assembly/disassembly can be triggered by an external or internal trigger (pH, temperature, concentration or enzyme activity). Thus, this methodology offers the possibility of incorporating different therapeutic agents during the process of self-assembly while also controlling their release through reversible structure transitions.

Among all the building blocks available for self-assembly, peptides have gained huge attention over the past decades due to their many advantageous properties. These include simple structure, facile synthesis, biodegradability, biocompatibility and customizable bioactivity. Diverse peptide-based self-assembling systems (peptide amphiphiles, cyclic peptides, surfactant-like peptides, aromatic dipeptides, dendritic peptides) [2] have been extensively studied to construct nanocarriers for the delivery of different therapeutics (oligonucleotides, hydrophobic drugs). The release of entrapped molecules could be easily triggered by protease hydrolysis of peptide bonds or by cues from the environment.

Peptide amphiphile (PA) molecules are typically composed of a hydrophobic alkyl tail and a hydrophilic peptide sequence [3]. The hydrophilic peptide sequence can be designed for specific interaction with other molecules and cells. For intracellular delivery, cell-penetrating peptides (CPPs), a family of short peptides that are able to translocate across cell membranes [4], have been widely used to facilitate cellular uptake of various therapeutic and imaging molecules, such as proteins and peptides, differentiating and constrast agents, DNA, siRNA, chemotherapeutic drugs and quantum dots [5]. However, the non-specificity of CPPs could lead to non-selective cytotoxicity and off-target side effects, resulting in a major limitation in systematic delivery. To overcome this problem, numerous approaches have been developed to selectively present CPPs at target sites [6]. In our work, multifunctional micelles are designed to display the cell-penetrating domain only

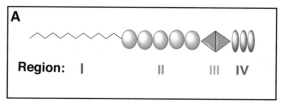

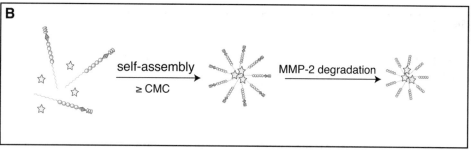

Fig. 1 (A) Rational design of multifunctional PAs for micelle self-assembly and targeted intracellular delivery; I: hydrophobic segment (e.g., $C_nH_{2n}O_2$, $14 < n < 20$); II: CPP sequence for mediating lipid membrane translocation (e.g., TAT peptide); III: enzyme cleavable peptide sequence (e.g., MMP-2 substrate GPX_1GILX_2G, where I denotes the expected cleavage site by MMP-2); IV: cell or tissue targeting peptide sequence (e.g., RGDS). **(B)** PA self-assembly into micellar structure with simultaneous drug (yellow star shape) encapsulation and presentation of CPP on the micelle surface triggered by enzymatic degradation

in situ where the enzyme matrix metalloproteinase 2 (MMP-2) is overexpressed. The multifunctional micelles are obtained upon self-assembly of rationally designed PAs, which consist of a 16-carbon alkyl tail (region I), a CPP domain (region II), a MMP-2 cleavable site (region III) and a tumor-targeting peptide sequence (region IV) (Fig. 1).

The methods to synthesize [7] and purify [8] PA molecules have already been reported in detail and will not be described in this chapter.

In this chapter, we describe the procedure to analyze the secondary structure of PAs using circular dichroism (CD), which also provides information on the type and stability of the assembled PA structures [9]. We also describe a method to determine the critical micelle concentration (CMC) using a solvatochromic fluorescence dye (Nile Red) [10], which provides information on the concentration that can trigger the PA self-assembly. Detailed procedures for using transmission electron microscopy (TEM) together with the negative staining technique are described to examine the shape and size of PA assemblies, which can strongly affect their blood circulation lifetime, bioavailability and cellular uptake [11].

The conditions for assessing the enzymatic degradation of PA micelles by MMP-2 in vitro are described, as well as methods to analyze the peptide fragments after degradation and calculate the degradation efficiency.

To assess the potential of the PA micelles as carriers for intracellular delivery, the protocol to conduct a sulforhodamine B (SRB) cytotoxicity assay [12] is described, as well as a qualitative evaluation of cell internalization of the PA micelles by confocal laser scanning microscopy (CLSM).

2   Materials

2.1   Preparation of PA Solution

1. PA (previously synthesized, purified and lyophilized) powder.
2. Milli-Q water.
3. 0.1 M Hydrochloric acid (HCl) solution.
4. 0.1 M Sodium hydroxide (NaOH) solution.
5. Milli-Q water purification system.
6. pH meter.
7. 0.22 μm Syringe filter.

2.2   Characterization of PA Secondary Structure

1. 1 mm Path length quartz cuvette.
2. Chirascan™ circular dichroism (CD) spectropolarimeter.

2.3   Determination of PA Critical Micelle Concentration (CMC)

1. 50 μM Nile Red acetone solution: Weigh out 0.50 mg Nile Red and place it in a 50 mL centrifuge tube. Add 31.41 mL acetone to the centrifuge tube and the final concentration of Nile Red solution should be 50 μM (*see* **Note 1**).
2. 0.5 mL Eppendorf tube.
3. Ethanol.
4. Vacuwash cuvette washer.
5. 10 mm Quartz cuvette.
6. Fluorolog® spectrofluorometer.

2.4   Morphological Characterization of Assembled PA Nanostructures

1. 2% (w/w) Uranyl acetate aqueous solution: Weight out 0.2 g of uranyl acetate (*see* **Note 2**) and add carefully into a 25 mL beaker. Measure 10 mL Milli-Q water and pour slowly into the beaker. Then stir the solution for at least 30 min. Filter the solution with a 0.22 μm syringe filter and then aliquot into 2 mL screw-cap brown bottle and label properly. The prepared 2% uranyl acetate solution should be kept at 4 °C in the dark place and filtered again before using.
2. Carbon film coated copper TEM grid (400 mesh).
3. PELCO easiGlow™ glow discharge unit.
4. Reverse (self-closing) TEM tweezer.
5. TEM grid storage box.

6. Filter paper.

7. Technai12 TWIN TEM with SIS Megaview III wide-angle CCD camera.

2.5 Evaluation of PA Degradation by MMP-2

1. TCNB buffer (50 mM Tris, 10 mM CaCl₂, 150 mM NaCl, and 0.05% Brij™ L23, pH 7.4): Dissolve 0.24 g Tris, 0.04 g CaCl₂, 0.35 g NaCl and 0.02 g Brij™L23 in 35 mL Milli-Q water. After all the reagents are completely dissolved, adjust the pH to 7.4 using 0.1 M HCl solution. Add Milli-Q water to a final volume of 40 mL.

2. MMP-2 (active, human, recombinant, CHO cells) (*see* **Note 3**).

3. 0.1 mL Eppendorf tube.

4. Water bath.

5. Liquid nitrogen.

6. Trifluoroacetic acid (TFA).

7. Acetonitrile (ACN).

8. Milli-Q water.

9. Waters® clear glass 12 × 32 mm screw neck Qsert vial, 300 μL volume

10. Alliance® high-performance liquid chromatography (HPLC) system with Waters® e2695 separations module, 2489 UV/Vis detector, Xbridge™ reverse phase column (C18, 4.6 × 150 mm, 3.5 μm), and Empower® chromatography data software.

11. Electrospray Ionization Mass Spectrometry (ESI-MS).

2.6 Evaluation of PA Cytotoxicity

1. Human fibrosarcoma HT-1080 cell line.

2. Human embryonic kidney 293T cell line.

3. 10% (w/v) Trichloroacetic acid (TCA) aqueous solution: Add 22 mL of Milli-Q water to a bottle containing 15 g of solid TCA to dissolve the TCA first and then continuously add Milli-Q water until the final volume is 150 mL. This will give 150 mL of a 10% (w/v) TCA solution.

4. Sulforhodamine B (SRB) solution (0.4% in 1% acetic acid) (*see* **Note 4**).

5. 1% (v/v) Acetic acid aqueous solution: Dilute the 10% (v/v) acetic acid solution tenfold with Milli-Q water.

6. 10 mM Tris base solution: Weight out 0.121 g Tris and dissolve in 90 mL Milli-Q water first and continuously add water to set the final volume of 100 mL.

7. 96-Well culture plate.

8. Sterile gauze.

9. SPECTROstar Nano® microplate reader.

2.7 Evaluation of PA
Micelle Cell Uptake

1. Human fibrosarcoma HT-1080 cell line.
2. Human embryonic kidney 293T cell line.
3. Coumarin-6.
4. Hexafluoroisopropanol (HFIP).
5. Dulbecco's phosphate-buffered saline (DPBS) solution.
6. Hanks' balanced salt solution.
7. 20 μM Lysotracker® Red staining solution: Dilute the Lysotracker® Red stock solution (1 mM) 50-fold with DPBS (*see* **Note 5**).
8. 2 mg/mL Hoechst® 33342 staining solution: Dilute the Hoechst® 33342 stock solution (10 mg/mL) fivefold with DPBS.
9. Collagen I coated chambered glass coverslip with 8 wells.
10. Leica TCS SP2 confocal laser scanning microscope (CLSM).

3 Methods

3.1 Characterization
of PA Secondary
Structure

1. Dissolve the PAs in Milli-Q water at a specific concentration (0.01–0.1 mM) (*see* **Note 6**) and adjust the pH to 7.4 using 0.1 M HCl or NaOH solution. The prepared samples are filtered through 0.22 μm syringe filters to remove any scattering particles before characterization.
2. Wash the cuvette (1 mm path length) with detergent first and then rinse with Milli-Q water. Use pure nitrogen to dry the cuvette.
3. Pipette 0.6 mL of Milli-Q water or prepared PA solution to the cleaned cuvette.
4. Record the spectrum on a CD spectropolarimeter (*see* **Note 7**) from 180 to 320 nm at 25 °C with a 10 nm/min scan rate, 1 s integration time, 1 nm bandwidth, and 3 spectral acquisitions.
5. Record first the baseline by measuring the cuvette containing only Milli-Q water and subtract this spectrum from each spectrum obtained for PA solution.
6. Convert the CD signals to the concentration-independent parameter of mean residue ellipticity $[\theta]_\lambda$ using the following equation:

$$[\theta]\lambda = \frac{[\theta]_{obs}}{10 \cdot c \cdot l \cdot n} \tag{1}$$

where $[\theta]_\lambda$ is the mean residue ellipticity at λ in deg·cm²·dmol⁻¹, $[\theta]_{obs}$ is the observed value of ellipticity from the instrument at λ in mdeg, c is the concentration of PAs in M, l is the light path

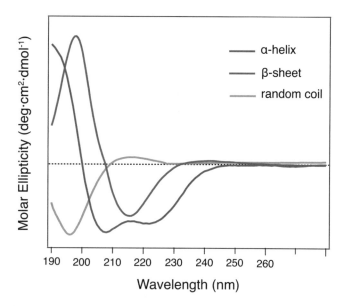

Fig. 2 Illustration of CD spectra showing three conformations typically observed in PA molecules: α-helix, β-sheet and random coil

Table 1
Positive and negative maxima in the CD spectra for typical PA conformations: α-helix, β-sheet, and random coil

| Conformation | Wavelength, λ (nm) | | |
	Positive maximum	Negative maximum	Zero
α-Helix	190–195	208 and 222	202 and 250
β-Sheet	195–197	217–218	207 and 250
Random coil	218	197 and 240	211, 234 and 250

length of the cuvette in cm, and n is the number of amino acid residues in the PA molecule.

7. Plot the mean residue ellipticity $[\theta]_\lambda$ as function of wavelength. Qualitatively compare the plotted spectrum with the spectra of the three typical peptide secondary structures, α-helix, β-sheet, and random coil (Fig. 2, Table 1), and assign the secondary structure of the PAs.

3.2 Determination of PA Critical Micelle Concentration (CMC)

1. Add 10 μL of Nile Red solution to a set number of 0.5 mL Eppendorf tubes. Leave all the Eppendorf tubes open in the dark fume hood and wait for about 5 h to allow the acetone to evaporate completely (*see* **Note 8**).

2. Prepare PA stock solution at 10 mM in Milli-Q water and adjust the pH to 7.4 using 0.1 M HCl or NaOH solution. Dilute the PA stock solution to a series of desired concentrations with Milli-Q water (e.g., 1, 0.8, 0.6, 0.4, 0.2, 0.1, 0.08, ..., 0.004, 0.002, 0.001 mM).

3. Add 0.5 mL of the diluted series of PA solutions to the Eppendorf tubes containing the dry Nile Red film. Prepare a blank sample by adding 0.5 mL Milli-Q water to the Eppendorf tube. The final concentration of Nile Red in each of the Eppendorf tubes should be 1 μM.

4. Leave all the samples in a dark place overnight to allow the solubilization equilibrium of Nile red.

5. Wash the cuvette (10 mm path length) in a Vacuwash cuvette washer with Milli-Q water first and then ethanol.

6. Pipette 0.5 mL of the prepared sample solution to the cleaned sample cuvette (*see* **Note 9**).

7. Record the fluorescence emission spectrum of each sample on a spectrofluorometer (*see* **Note 10**) ranging from 570 to 700 nm, with an excitation wavelength at 550 nm. The bandwidths of both emission and excitation are kept at 5 nm and the spectral acquisition number is set at 3.

8. Determine the maximum intensity and the corresponded wavelength of each spectrum after averaging the three acquisition data of each sample and subtracting the average acquisition data of the blank sample.

9. Plot both the maximum fluorescence intensity and the corresponding wavelength as function of logarithm of PA concentration. The CMC can be determined at the point where there is a sharp increase in the fluorescence intensity (Fig. 3a) and a hypsochromic effect (blue shift) (Fig. 3b).

3.3 Morphological Characterization of Assembled PA Nanostructures

1. Prepare the PA solution at a specific concentration (≥CMC) in Milli-Q water and adjust the pH to 7.4 using 0.1 M HCl or NaOH solution. Age the PA solution overnight (*see* **Note 11**) and filter through 0.22 μm syringe filter before the next step.

2. Hold the TEM grid (*see* **Note 12**) in a reverse tweezer.

3. Place a drop (~10 μL) of the PA solution onto the carbon film coated copper grid and remove the excess of PA solution after 1 min by touching the rim of the grid with a piece of filter paper.

4. Apply a drop (~10 μL) of the prepared uranyl acetate solution onto the grid, wait for 30 s for the negative staining, and remove the staining solution by touching the rim of the grid with a piece of filter paper.

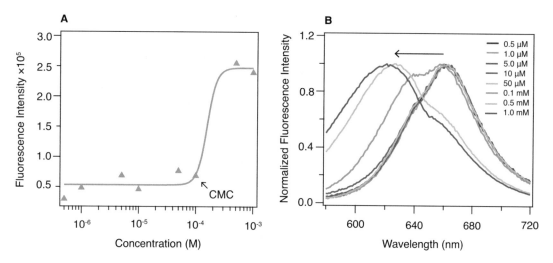

Fig. 3 (**A**) Plot of maximum fluorescence intensity of Nile Red versus logarithm of PA concentration; CMC is marked at 0.1 mM in the graph. (**B**) Emission spectra of Nile Red upon incubation with PAs at a different concentrations (all the spectra in the graph are normalized by the maximum emission); blue shift is indicated by arrow

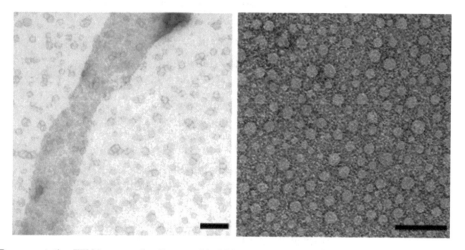

Fig. 4 Representative TEM images of self-assembled PA micelles negatively stained with uranyl acetate; scale bars: 100 nm

5. Allow the grid to dry completely at room temperature for at least 3 h and store the grid in the TEM grid box.

6. Examine the assembled PA nanostructures using the TEM operating at an acceleration voltage of 100 kV and take images using the wide-angle CCD camera.

7. Representative assembled PA micelles are shown in Fig. 4.

3.4 Evaluation of PA Degradation by MMP-2

1. Dissolve PAs and MMP-2 in the same container using an adequate volume of TCNB buffer to give the desired concentration. The volume of the TCNB buffer added should consider the number of time intervals to be analysed. The concentration of PAs should be in the range that guarantees the capture of well-defined micelle nanostructures under TEM. The typical concentration of MMP-2 used is 20 nM, as reported in the literature [13].

2. Control sample is prepared following the same step as described above by dissolving only the PAs in TCNB buffer without MMP-2.

3. Incubate the mixed solution in a water bath at 37 °C.

4. Transfer 50 μL of the mixed solution to a 0.1 mL Eppendorf tube at desired time intervals (e.g., 0, 0.5, 1, 2, 4, ..., 12, 24 h) and freeze it immediately in liquid nitrogen.

5. Defrost the sample and transfer quickly to the Qsert vial. The Qsert vial is then put into the sample acquisition panel of the HPLC system for injection. A Qsert vial containing only 50 μL TCNB buffer is prepared as blank and injected first.

6. The HPLC system is programmed to perform a run under the following conditions:

   Injection volume: 30 μL.

   Flow rate: 1.0 mL/min.

   Mobile phase: (A): 0.1% TFA (v/v) in H_2O.

   (B): 0.1% TFA (v/v) in ACN.

   Gradient: 2–98% of (B) in 25 min.

   UV detection: 220 nm

7. Collect the fractions when significant UV signal appears and store them for later mass identification by ESI-MS analysis.

8. Integrate the area (A) of the peaks in the HPLC chromatogram (Fig. 5) using the Empower® chromatography data software and calculate the enzyme degradation efficiency using the following equation:

$$\text{Enzyme degradation effiency }(\%) = \frac{A_{[A]}}{A_{[A]} + A_{[B]} + A_{[PA]}} \times 100\% \quad (2)$$

where $A_{[PA]}$ represents the area corresponding to the PA peak (uncleaved), while $A_{[A]}$ and $A_{[B]}$ represent the peak areas of the peptide fragments resulting from the enzymatic cleavage (Fig. 5).

3.5 Evaluation of PA Cytotoxicity

All the cell culturing work must be performed under sterile cell culture conditions: humidified air, 5% CO_2 and 37 °C.

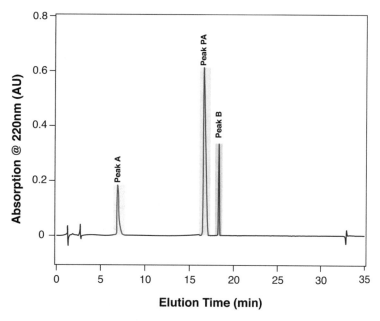

Fig. 5 Representative HPLC chromatogram of the peptide mixture obtained after PA incubation with MMP-2 for 2 h (*see* Subheading 3.4, **step 3**). HPLC running conditions were as described in Subheading 3.4, **step 6**, and peptide fragments resulting from MMP-2 degradation are identified as Peak A and Peak B. Integration of the peak areas shown in the chromatogram is used to calculate the MMP-2 degradation efficiency for the given incubation time

1. Seed 1×10^4 cells (*see* **Note 13**) in 200 μL growth medium per well of the culture plate and allow 24 h for cell attachment. A blank 8-well column, containing only the growth medium, is set to determine the blank background absorption (*see* Fig. 6).

2. Remove the growth medium from each well by gently flipping the culture plate upside down on a piece of gauze.

3. Add 200 μL of the fresh growth medium containing different concentration of PAs to corresponding columns of the culture plate. The PA concentration can be set from 1 nM to 0.1 M. We suggest log or half-log dilutions to cover a broad concentration range and to ensure that the final data points on the growth curve are homogeneously distributed. Pipette PA-free growth medium into both the blank and control 8-well column, respectively (*see* Fig. 6).

4. Incubate cells during a desired PA exposure time, which can be from 2 to 96 h. We usually set the PA exposure time as 48 h.

5. Gently remove the growth medium by flipping the culture plate and remove the residual medium by keeping the culture plate upside down on a piece of gauze (*see* **Note 14**).

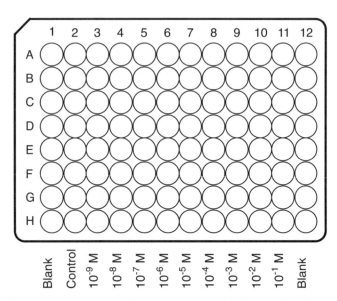

Fig. 6 Representative schematic of a 96-well cell culturing plate for cytotoxicity evaluation, which shows different columns for blank, control and PA concentrations

6. Fix the cells by gently adding 100 μL of the cold 10% TCA solution (4 °C) into each well of the culture plate and incubate for at least 1 h at 4 °C.

7. Rinse the culture plate with Milli-Q water (200 μL/well), then quickly flip the culture plate over a sink and remove roughly the residues onto a piece of gauze. Repeat five times to remove completely the TCA solution, serum protein, etc.

8. Air-dry the culture plate for at least 2 h until no moisture is visible.

9. Pipet 30 μL of the 0.4% SRB solution into each well of the culture plate using a multichannel pipette and allow cells to be stained for 30 min in a dark place.

10. Remove the staining solution and rinse the culture plate quickly with 1% acetic acid solution (200 μL/well) three times (*see* **Note 15**). The culture plate is air-dried again for at least 2 h until no moisture is visible.

11. Solubilize the protein-bound SRB by adding 10 mM Tris solution (100 μL/well). Pipette the Tris solution up and down and wait for at least 10 min to allow enhanced mixing of the SRB.

12. Measure the absorbance using an automated plate reader at a wavelength of 565 nm (*see* **Note 16**).

13. The percentage of cell viability can be calculated using the following equation:

$$\text{Cell viability}\,(\%) = \frac{\text{Abs}\left[\text{Sample}\right] - \text{Abs}\left[\text{Blank}\right]}{\text{Abs}\left[\text{Control}\right] - \text{Abs}\left[\text{Blank}\right]} \times 100\% \qquad (3)$$

All the data should be averaged and the standard deviation (SD) calculated (*see* **Note 17**).

3.6 Evaluation of PA Micelle Cell Uptake

1. Dissolve both the PA and Coumarin-6 in HFIP at a molar ratio of 1000:1 and dry under vacuum to generate a thin film by rotary evaporation. Control sample is prepared containing only Coumarin-6 without the PAs.

2. Redissolve the thin film by adding a desired volume of sterilized DPBS and age overnight in a dark place. The final concentration of Coumarin-6 should be 30 µM.

3. Mix the prepared PA solutions with a desired volume of fresh growth medium and keep the final concentration of Coumarin-6 at 0.3 µM.

4. Seed the HT-1080 and 293T cells onto the collagen I coated chambered coverslip at 4×10^4 cells/well and allow cell attachment overnight.

5. Remove the previous medium and add growth medium containing PA/Coumarin-6 into the culture plate (200 µL/well) and then incubate for 4 h.

6. Add 1 µL Lysotracker® Red and 1 µL Hoechst® 33342 directly into the growth medium of each well at a final concentration of 100 nM and 10 µg/mL, respectively, and wait for 30 min for the lysosome and nucleus staining.

7. Remove the growth medium and wash the cells twice with Hanks' balanced salt solution.

8. Add fresh growth medium without phenol red into the culture plate (200 µL/well).

9. Qualitatively evaluate the cell uptake by imaging the cells in a CLSM. Different filters are used to record the fluorescence from Hoechst® 33342 (excitation/emission: 350 nm/461 nm; blue fluorescence), Coumarin-6 (excitation/emission: 443 nm/505 nm; green fluorescence), and Lysotracker® red (excitation/emission: 577 nm/590 nm; red florescence).

4   Notes

1. The centrifuge tube containing the Nile Red acetone solution should be sealed with Parafilm and kept in a dark place before use. It can be stored at 2–8 °C for up to 2 weeks.

2. Uranyl acetate is toxic and radioactive (alpha emitter). Inhalation is the most serious route of exposure. All uranyl acetate solids and solutions should be handled in a fume hood labeled with a caution radioactive sticker and all solid and solution wastes containing uranyl acetate need to be disposed of as radioactive waste and disposed of properly.

3. The MMP-2 should be aliquoted and frozen under −80 °C after initial use. Repeated freeze/thaw cycle should be avoided to keep the activity of MMP-2.

4. The 0.4% SRB solution should be stored at 4 °C and protected from light. It should be filtered using a syringe filter before use, if precipitated SRB crystals appear.

5. The Lysotracker® Red and Hoechst® 33342 dyes are known mutagen and should be handled with care. After dilution in DPBS, the staining solutions should be stored in single-use aliquot, frozen under −20 °C and protected from light.

6. The concentration of PA solution needs to be slightly adjusted according to different samples to get significant and smooth signal while preventing saturation of the detector.

7. The CD spectropolarimeter should be purged with pure nitrogen for at least 30 min at the flow rate suggested by the manufacturer before use. This will help to create an oxygen-free environment and prevent the generation of ozone by far-UV radiation.

8. The environmental conditions (e.g., pH, temperature, ionic concentration and solvent) can significantly affect the noncovalent interactions of PAs and thus influence their self-assembly process. The final concentration of acetone should be kept as low as possible to prevent disrupting the PA assemblies.

9. The sample solution could only be gently vortexed before transferring to the spectrofluorometer cuvette. Vigorous vortex-ing or sonication may destroy the assembled nanostructures.

10. The lamp of the spectrofluorometer should be switched on 30 min before starting data collection to warm up and stabilize the excitation light.

11. PA self-assembly is instaneous at concentration higher than CMC. However, from our experience, aging the PA solution overnight typically allows observing well-defined nanostructures under TEM.

12. The carbon film coated on the TEM grid has the tendency to adsorb hydrocarbons and be hydrophobic. To get clean TEM grids with consistent quality, a glow discharge treatment is needed to clean the grids and make the carbon film surface hydrophilic, which helps the aqueous solution to spread on the grid easily.

13. The cell density is chosen based on prior cell growth kinetics study to ensure the exponential cell growth during the entire SRB assay and the final absorption intensity in the range of 1.5–2.0 since the curve of absorption intensity versus dye concentration is linear only in a limited range.

14. For some cell lines having poor attachment to the culture plate, the removal of growth medium before the TCA fixation step might not be suitable.

15. The wash time should be kept to a minimum to reduce desorption of protein-bound dye when using 1% acetic acid solution.

16. If high absorption intensity (>2.0 units) is observed, it is necessary to use a suboptimal wavelength (490–550 nm) to facilitate the absorbance reading of wells.

17. At least three independent experiments should be conducted and the SD value should not exceed 20% since the SRB assay is highly reproducible. SD value of all the wells in a column should also not exceed 20% to ensure homogeneous cell distribution in the plate.

Acknowledgments

Y. Shi thanks China Scholarship Council for her PhD Scholarship (No. 201307060020) and Queen Mary University of London for the Postgraduate Research Fund. H. S. Azevedo acknowledges the financial support of the European Union under the Marie Curie Career Integration Grant SuprHApolymers (PCIG14-GA-2013-631871).

References

1. Torchilin VP (2014) Multifunctional, stimuli-sensitive nanoparticulate systems for drug delivery. Nat Rev Drug Discov 13(11):813–827. https://doi.org/10.1038/nrd4333

2. Mendes AC, Baran ET, Reis RL, Azevedo HS (2013) Self-assembly in nature: using the principles of nature to create complex nano-biomaterials. Wiley Interdiscip Rev Nanomed Nanobiotechnol 5(6):582–612. https://doi.org/10.1002/wnan.1238

3. Cui H, Webber MJ, Stupp SI (2010) Self-assembly of peptide amphiphiles: from molecules to nanostructures to biomaterials. Biopolymers 94(1):1–18. https://doi.org/10.1002/bip.21328

4. Copolovici DM, Langel K, Eriste E, Langel U (2014) Cell-penetrating peptides: design, synthesis, and applications. ACS Nano 8(3):1972–1994. https://doi.org/10.1021/nn4057269

5. Juliano RL, Alam R, Dixit V, Kang HM (2009) Cell-targeting and cell-penetrating peptides for delivery of therapeutic and imaging agents. Wiley Interdiscip Rev Nanomed Nanobiotechnol 1(3):324–335. https://doi.org/10.1002/wnan.4

6. MacEwan SR, Chilkoti A (2013) Harnessing the power of cell-penetrating peptides: activatable carriers for targeting systemic delivery of cancer therapeutics and imaging agents. Wiley Interdiscip Rev Nanomed Nanobiotechnol 5(1):31–48. https://doi.org/10.1002/wnan.1197

7. Mata A, Palmer L, Tejeda-Montes E, Stupp SI (2012) Design of biomolecules for nanoengineered biomaterials for regenerative medicine. Methods Mol Biol 811:39–49. https://doi.org/10.1007/978-1-61779-388-2_3

8. Conlon JM (2007) Purification of naturally occurring peptides by reversed-phase

HPLC. Nat Protoc 2(1):191–197. https://doi.org/10.1038/nprot.2006.437

9. Trent A, Marullo R, Lin B, Black M, Tirrell M (2011) Structural properties of soluble peptide amphiphile micelles. Soft Matter 7(20):9572–9582. https://doi.org/10.1039/c1sm05862b

10. Stuart MCA, van de Pas JC, Engberts J (2005) The use of Nile red to monitor the aggregation behavior in ternary surfactant-water-organic solvent systems. J Phys Org Chem 18(9):929–934. https://doi.org/10.1002/poc.919

11. Goldberg M, Langer R, Jia X (2007) Nanostructured materials for applications in drug delivery and tissue engineering. J Biomater Sci Polym Ed 18(3):241–268

12. Skehan P, Storeng R, Scudiero D, Monks A, McMahon J, Vistica D, Warren JT, Bokesch H, Kenney S, Boyd MR (1990) New colorimetric cytotoxicity assay for anticancer-drug screening. J Natl Cancer Inst 82(13):1107–1112

13. Patterson J, Hubbell JA (2010) Enhanced proteolytic degradation of molecularly engineered PEG hydrogels in response to MMP-1 and MMP-2. Biomaterials 31(30):7836–7845. https://doi.org/10.1016/j.biomaterials.2010.06.061

Chapter 3

Electrospinning Functionalized Polymers for Use as Tissue Engineering Scaffolds

Lesley W. Chow

Abstract

Electrospinning polymers is a versatile technique to generate fibrous, three-dimensional scaffolds for tissue engineering applications. Modifying polymers with functional groups prior to electrospinning offers the opportunity to control the spatial presentation of functional groups within the scaffold as well as incorporate multiple bioactive cues. This chapter describes methods to modify poly(ε-caprolactone) (PCL) with peptides and electrospin these peptide-PCL conjugates to functionalize a scaffold surface in a single step. Methods to adapt standard electrospinning setups to create single- or dual-peptide gradients within a single construct are also described.

Key words Tissue engineering, Regenerative medicine, Electrospinning, Functionalized polymers, Biodegradable scaffold, Gradient biomaterials

1 Introduction

Tissue engineering is a multidisciplinary approach that combines cells, biomaterials, and biological factors to repair or replace tissue and organ functions [1]. Scaffolds for tissue engineering are typically designed to mimic the extracellular matrix (ECM) that surrounds cells in native tissues. Electrospinning is a popular and versatile technique to fabricate fibrous, three-dimensional scaffolds that resemble the ECM using a variety of polymers of natural or synthetic origins [2–6]. The typical setup includes a syringe pump, a high-voltage source, and a collector for the resulting electrospun fibers. The syringe pump provides constant flow of a polymer solution from a needle while an electric field is applied using the high-voltage source. The polymer solution at the needle tip becomes electrically charged, forming a suspended conical droplet until the applied electric field is strong enough to overcome the surface tension of the solution. A tiny jet emerges from the surface of the droplet and is drawn toward the collector where the resulting electrospun fibers form a nonwoven mesh. The processing parameters

Kanika Chawla (ed.), *Biomaterials for Tissue Engineering: Methods and Protocols*, Methods in Molecular Biology, vol. 1758, https://doi.org/10.1007/978-1-4939-7741-3_3, © Springer Science+Business Media, LLC, part of Springer Nature 2018

can be tailored to generate desired fiber diameters and orientations as well as overall scaffold dimensions [4, 7].

Electrospinning with natural ECM polymers such as hyaluronic acid and collagen offers good biocompatibility and intrinsic bioactivity, such as growth factor binding and cell adhesion cues, compared with using synthetic polymers but is limited by batch-to-batch variability and poor mechanical properties [8]. The electrospinning process, which often involves organic solvents in addition to the high voltages, can also denature these biomolecules and affect their bioactivity.

Synthetic polymers offer several advantages over natural polymers because they are readily available and easily processed for more reproducible results. One major limitation with using synthetic polymers, however, is that the surface chemistry of the resulting electrospun scaffold is not ideal for cell adhesion. The surface can be chemically and physically modified to display bioactive molecules and ligands to enhance cell-material interactions for biomedical applications [9]. This traditionally involves post-fabrication functionalization techniques, such as plasma treatment, covalent grafting, or physisorption, to introduce the desired functionality, but these approaches add many processing steps and may compromise bulk scaffold properties [9, 10]. In addition, it is difficult to control the concentration and orientation of the biomolecule on the surface since attachment typically involves nonspecific reactions that can occur with multiple sites on the biomolecule. An alternative approach is to modify the polymer with functional groups prior to electrospinning [10–17]. During electrospinning, the functional segment of the modified polymer is driven to the fiber surface to functionalize and fabricate the scaffold in a single step. This strategy can also be combined with advanced electrospinning techniques to control both the concentration and spatial presentation of functional groups within the scaffold [14–16]. During tissue development and remodeling, gradients of chemical cues regulate cell behavior such as migration, proliferation, and differentiation [18]. For example, cell adhesion peptides can be presented in a concentration gradient through a scaffold to guide cell migration.

This chapter focuses on methods to modify poly(ε-caprolactone) (PCL), a biocompatible and biodegradable polymer that is widely used for tissue engineering scaffolds, with specific peptide sequences to form peptide-PCL conjugates and adapt standard electrospinning setups to create gradients of one or more peptides within a single scaffold [15, 16]. The protocol describes the step-by-step procedure to synthesize the peptide-PCL conjugates using commercially available PCL and a versatile peptide design. The conjugates are added to unmodified, high-molecular-weight PCL, which acts as the carrier polymer for electrospinning to ensure adequate fiber formation and maintain bulk scaffold properties. The electrospinning technique has been described extensively in the literature and in other chapters from this series [2–4] and therefore will not be discussed at length here. Instead, methods to utilize two programmable syringe pumps with a standard electrospinning setup to create gradients of peptides within a scaffold are described. This platform can be broadly applied with other functional groups and polymers to cover a wide range of tissue engineering applications. For example, similar strategies have been used to conjugate PCL with other functional groups such as carbohydrates [12] or initiating groups for controlled radical polymerization [16]. Other polymers such as poly(ethylene oxide) (PEO) [10] and poly(lactic-co-glycolic acid) (PLGA) [11] have also been modified with peptides to form functionalized polymers for electrospinning. The protocol described herein therefore addresses a versatile strategy to generate multifunctional scaffolds for tissue engineering.

2 Materials and Equipment

2.1 Peptide-PCL Conjugate Synthesis

2.0.1 Materials and Supplies

1. Polymer to be modified: Poly(ε-caprolactone) (PCL), average M_w 14,000, average M_n 10,000, α,ω-dihydroxy functional polymer (Sigma-Aldrich, USA) (*see* **Note 1**).

2. Heterobifunctional linker: p-Maleimidophenyl isocyanate (PMPI) (*see* **Note 2**).

3. Solvent: N-methyl-2-pyrrolidone (NMP), anhydrous (*see* **Note 3**).

4. Glass vials: 14 mL capacity with screw cap and foil insert (to enhance seal).

5. Peptides: Sequences should include (1) a terminal cysteine; (2) a spacer, such as three glycines; and (3) a bioactive sequence of interest (*see* **Note 4**).

6. Syringes: Solvent resistant, 10 mL capacity, Luer-lock tip.

7. Needles for extracting solvent: 18-gauge, 6.0 in. length, stainless steel, threaded cap for Luer-Lock syringe.

8. Glass round-bottom flask, 50 mL.

9. Rubber stopper with sleeves to seal round-bottom flask.

10. Magnetic stir bar.

11. Magnetic stirring plate.

12. Nitrogen gas (N_2).

13. Aluminum foil.

14. Diethyl ether (DEE).

15. Ultrapure water.

16. Centrifuge tubes, 50 mL capacity.

17. Glass filtering flask, 500 mL with side arm.

18. Glass filtration Buchner funnel, 150 mL, fritted, G3 medium porosity.

19. Rubber adapter for Buchner funnel.

20. Metal spatula.

21. Nuclear magnetic resonance (NMR) tubes with dimensions suitable for available NMR instrument.

22. Deuterated solvents: DCM-d2 and DMSO-d6.

23. Glass Pasteur pipets.

24. Pipet bulb.

2.1.1   Equipment

1. Sonicator.

2. Centrifuge capable of spinning down samples in 50 mL tubes at 5500–7500 rcf.

3. Vacuum pump.

2.2   Electrospinning

2.2.1   Materials and Supplies

1. Solvent: 1,1,1,3,3,3-Hexafluoro-2-propanol (HFIP) (*see* **Note 5**).

2. Carrier polymer for electrospinning: Poly(ε-caprolactone) (PCL), average M_n 80,000 (Sigma-Aldrich, USA) (*see* **Note 6**).

3. Glass vials: 14 mL capacity with phenolic screw cap and foil insert (to enhance seal).

4. Parafilm to seal vials.

5. Needle: 18-gauge, 1.5 in. length, blunt end, stainless steel needle, threaded cap for Luer-Lock syringe.

6. Syringes: Solvent resistant, 10 mL capacity, Luer-lock tip.

7. Tubing: Polytetrafluoroethylene (PTFE), 0.25 in. (inner diameter), 0.3125 in. outer diameter (*see* **Note 7**).

8. 3-Way stopcock: Polyvinylidene fluoride (PVDF) stopcock with polypropylene insert, female Luer-Lock (Cole-Parmer, USA).

9. Male Luer-Lock adapters: Polycarbonate or polypropylene (PP), male Luer to 0.25 in. barb adapter (Cole-Parmer, USA).

10. Female Luer-Lock adapters: Polycarbonate or polypropylene (PP), female Luer to 0.25 in. barb adapter (Cole-Parmer, USA).

11. Aluminum foil to place on grounded target to collect electrospun fibers.

2.2.2   Equipment

1. Roller mixer that can continuously mix solutions in the glass vials.

2. Two programmable high-pressure syringe pumps (example model: Aladdin 1010, World Precision Instruments, UK).

3. Electrospinning setup: Voltage supply, lab jack, grounded target (as described previously in other chapters from the *Methods in Molecular Biology* series [2–4]).

3   Methods

3.1   PCL-Maleimide (PCL-Mal) Synthesis

The following protocol corresponds to the synthesis steps shown in Fig. 1. All reactions should be completed in a chemical fume hood.

1. Add PCL (M_w 14,000) to round-bottom flask with magnetic stir bar and seal with sleeved stopper (*see* **Note 8**).

2. Purge flask for 10–20 min by pumping in N_2 through a needle through the sleeved stopper with another needle to allow air to escape from the flask (*see* **Note 9**).

3. Add anhydrous NMP using a syringe via needle through the stopper to dissolve PCL at 60 mg/mL and stir on magnetic stirring plate until fully dissolved. Sonication may be required to ensure that the polymer is fully dissolved (*see* **Note 10**).

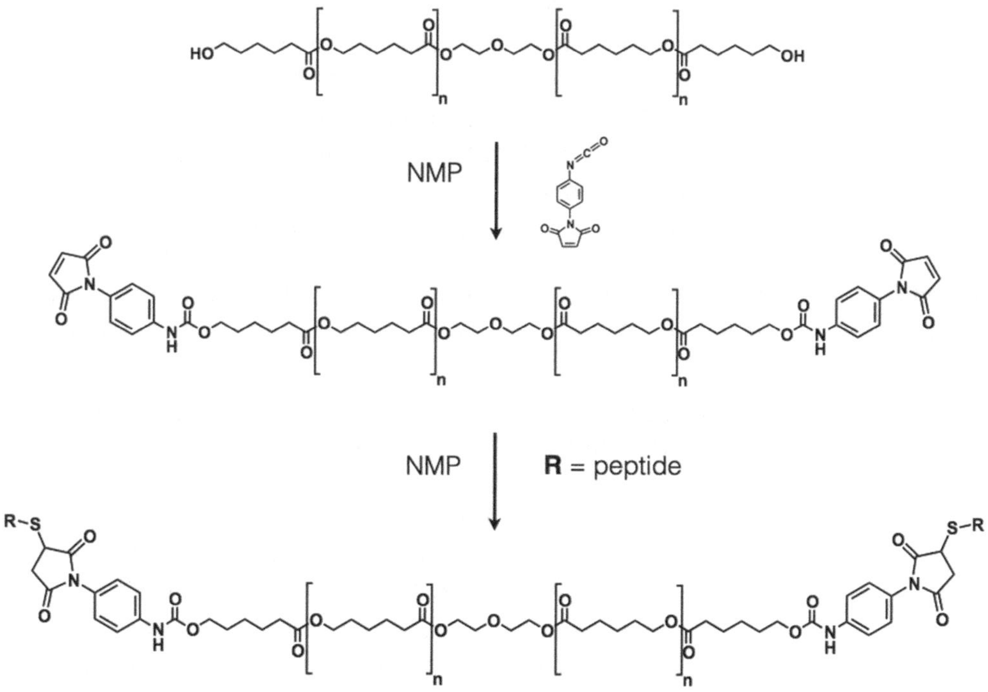

Fig. 1 Synthesis scheme of peptide-PCL conjugate. PCL diol (1) is modified with p-maleimidophenyl isocyanate (PMPI) (2) to form PCL-mal (3). The sulfhydryl on the cysteine of the peptide (4) is reacted with the maleimide on the PCL-mal to form the peptide-PCL conjugate (5)

4. In a glass vial, dissolve PMPI in 1 mL of anhydrous NMP at 20 molar equivalents to the PCL. Additional NMP may be added depending on PMPI solubility and reaction scale (*see* **Note 11**).

5. Add PMPI solution dropwise using a syringe via needle through the stopper to the PCL solution while stirring.

6. Cover the flask with aluminum foil to protect from light and allow the reaction to continue overnight to form PCL modified with maleimide (PCL-mal).

7. Precipitate PCL-mal by adding the reaction solution dropwise using glass Pasteur pipet to DEE in a 50 mL tube. Use a tenfold volume of DEE compared to the PCL-mal reaction solution to allow adequate precipitation. For example, 4 mL of the PCL-mal reaction solution should be precipitated with at least 36 mL of DEE.

8. Centrifuge the precipitate at 5500–7500 rcf for 5 min and pour off supernatant solution (*see* **Note 12**).

9. Add fresh DEE to PCL-mal and repeat **step 8**.

10. Transfer PCL-mal to a fritted Buchner funnel and attach funnel to the filtering flask with the rubber adapter.

11. Add fresh DEE to the precipitate in the funnel and apply a vacuum to dry the PCL-mal.

12. Use the spatula to break up the precipitate and continue adding DEE and drying under vacuum until PCL-mal has a powder-like consistency (*see* **Note 13**).

13. To verify modification, dissolve 5 mg of PCL-mal in 700 μL of DCM-d2 in an NMR tube and analyze the sample using ^{1}H NMR (*see* **Note 14**).

14. Store PCL-mal in a −20 °C freezer until ready to use.

3.2 Peptide-PCL Conjugate Synthesis

The following protocol corresponds to the synthesis steps shown in Fig. 1. All reactions should be completed in a chemical fume hood.

1. Once PCL-mal modification has been verified, redissolve PCL-mal in anhydrous NMP and purge with N_2 as described for PCL in Subheading 3.1, **steps 1–3**.

2. Dissolve peptide in 1 mL of anhydrous NMP at 8 molar equivalents to the PCL. Additional NMP may be added depending on peptide solubility and reaction scale (*see* **Note 15**).

3. Add peptide solution dropwise using a syringe via needle through the stopper to the PCL-mal solution while stirring.

4. Allow reaction to continue overnight while stirring to form peptide-PCL conjugates.

5. Precipitate peptide-PCL conjugate following Subheading 3.1, **steps 7–9** (*see* **Note 16**).

6. Add ultrapure water to the peptide-PCL conjugate precipitate in the 50 mL tube and sonicate for 10 min to dissolve the excess peptide (*see* **Note 17**).

7. Transfer the peptide-PCL conjugate to a fritted Buchner funnel and apply a vacuum to remove the aqueous solution. Similar to previous steps, use the spatula to break up the precipitate and continue washing with ultrapure water. Follow Subheading 3.1, **steps 10–12**, to dry the precipitate with DEE.

8. To verify conjugation of the peptide to the PCL-mal, add 5 mg of the peptide-PCL conjugate to an NMR tube, dissolve in 700 µL of DMSO-d6, and then analyze the sample using ^{1}H NMR (*see* **Note 18**).

9. Store peptide-PCL conjugate in a −20 °C freezer until ready to use.

3.3 Electrospinning Scaffolds with a Peptide Gradient

The following protocol describes methods to create a gradient in peptide concentration by sequentially electrospinning unmodified PCL with increasing concentrations of the peptide-PCL conjugate using separate, programmable syringe pumps. All sample preparation and electrospinning should be completed in a chemical fume hood.

1. Add peptide-PCL conjugate at the desired concentration to unmodified PCL (M_n 80 K) and dissolve all molecules in HFIP in a glass vial. A typical concentration of unmodified PCL in HFIP is 12% w/v. Prepare a solution with only unmodified PCL in a separate glass vial. Seal both vials tightly with parafilm and mix on roller mixer overnight (*see* **Note 19**).

2. Fill the syringes with either the unmodified PCL solution (solution 1) or the conjugate solution (solution 2). Remove all air bubbles from the solution in the syringe (*see* **Note 20**).

3. The syringe pumps and electrospinning solutions should be set up as shown in Fig. 2. Briefly, attach a female luer adapter to the syringe with solution 1 and attach the tubing to the hose barb end. Attach the opposite end of the tubing to the hose barb end of a male luer adapter and connect the luer side of the adapter to the 3-way stopcock. Connect the syringe with solution 2 to the 3-way stopcock and attach a blunt-end needle to the remaining opening on the 3-way stopcock.

4. Position the 3-way stopcock so that only the syringe with solution 2 can pass through to the needle and the path for solution 1 is closed. Push the solution slowly through the needle until it is expelled (*see* **Note 21**).

5. Open the 3-way stopcock so that all valves are open. Push solution 2 through the tubing and needle until it is expelled (*see* **Note 22**).

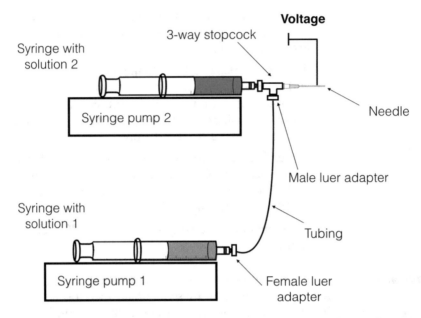

Fig. 2 Schematic of a dual-syringe pump setup for sequential electrospinning of two different polymer solutions. The syringes are connected using luer adapters, tubing, and a 3-way stopcock, and the solutions are pumped at opposing flow rates during electrospinning to control gradient formation

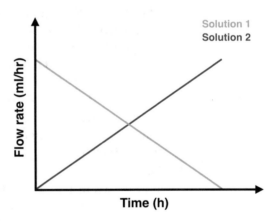

Fig. 3 Example of syringe pump flow rates to produce functional gradients within electrospun scaffold

6. Program each syringe pump according to the manufacturer's instructions to increase or decrease the flow rate as desired. For example, syringe pump 1 can be programmed with an initial flow rate of 2.0 mL/h and then gradually reduced to 0 mL/h while syringe pump 2 increases from 0 to 2.0 mL/h, as shown in Fig. 3 (*see* **Note 23**).

7. Attach the voltage source to the tip of the needle and electrospin on to foil according to preferred parameters as described in previous chapters [2–4].

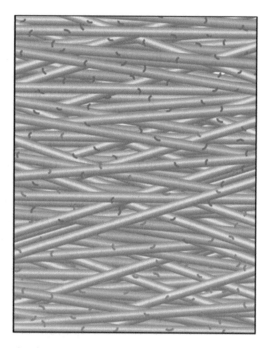

Fig. 4 Schematic of cross-sectional view of scaffold fabricated by sequentially electrospinning with two different peptide-PCL conjugates to form opposing gradients of conjugate 1 peptide (orange) and conjugate 2 peptide (blue) presented on the fiber surfaces

3.4 Electrospinning Scaffolds with Dual-Peptide Gradients

The methods described in Subheading 3.3 can be adapted easily with two different peptide-PCL conjugates by replacing the unmodified PCL solution with one that includes a peptide-PCL conjugate with a different amino acid sequence. The solutions containing two different peptide-PCL conjugates can be sequentially electrospun to create opposing peptide gradients as shown schematically in Fig. 4. Dual-peptide gradients can be used to present multiple spatial chemical cues, such as binding different biomolecules in opposing gradients within the scaffold, to guide cell behavior. This strategy can be tailored by manipulating the syringe pump program as desired and can be further adapted to incorporate more than two solutions for more complex scaffold designs.

4   Notes

1. The polymer to be modified should have a relatively low molecular weight (M_w) compared to the M_w (e.g., PCL with average M_n 80,000) typically used for electrospinning. The lower M_w reduces the signal intensity from the polymer so that the maleimide and peptide signals can be resolved accurately. This is critical to quantify the degree of modification.

2. The PMPI linker contains an isocyanate group that is reactive with the hydroxyl groups at the terminal ends of the PCL, forming a carbamate linkage. The isocyanate group is reactive with hydroxyl-containing reagents, including water, and must be protected from moisture and light. Store PMPI under nitrogen or argon at 4 °C.

3. The maleimide group on the PMPI linker can react with amines so the solvent must not contain primary amines. Anhydrous solvents are also necessary to protect the isocyanate group, which can react with water.

4. The peptide sequence must include (1) a cysteine, (2) a spacer, and (3) a bioactive sequence of interest. The cysteine provides a thiol used to attach the peptide to the PCL-maleimide via Michael addition [19]. The maleimide can also react with amines depending on the reaction conditions but reacts significantly faster and more efficiently with the thiol under the conditions described in this protocol. Peptides can therefore include amines in their sequence since the cysteine is available for selective conjugation. The spacer, such as multiple glycines, ensures that the bioactive peptide is fully presented on the fiber surface and not embedded within the bulk electrospun fiber, which may inhibit or limit activity. The bioactive peptide can also be modified with an orthogonal group, such as biotin for labeling with streptavidin-functionalized markers, to visualize the spatial location and density of the peptides [15].

5. Although other solvents such as chloroform and dichloromethane are commonly used for electrospinning PCL, HFIP is necessary to dissolve peptide-PCL conjugates.

6. Electrospinning typically requires polymers with high M_w to form homogeneous fiber morphologies. Unmodified PCL 80 K is therefore used as the bulk carrier polymer for the conjugates, which are added at varying concentrations to mediate the presentation density of peptides. The PCL block anchors the conjugate to the bulk electrospun fiber while the peptides are polarized to the fiber surface.

7. The tubing must be made of material that can withstand a large range of solvents. PTFE provides a flexible material that is resistant to many organic solvents used for electrospinning. Similarly, the 3-way stopcock and adapters must also be made of solvent-resistant materials, such as PVDF and PP.

8. The initial amount of PCL directly affects the total amount of peptide-PCL conjugate that can be produced. Calculations should be based on how much PMPI and peptide are available and how much conjugate is desired. For example, 100 mg PCL (M_w 14,000) requires a minimum of 30.6 mg PMPI for 20 molar equivalents to form the PCL-mal. The final amount

of conjugate produced is also affected by the PCL-mal product yield, which is typically around 80–85%. Larger amounts of PCL-mal should be synthesized to accommodate the loss.

9. The purging time should be increased if using a larger round-bottom flask.

10. The exact amount of PCL and NMP used is not crucial, particularly considering that the reactions are done with significant excess. However, the total volume should be minimized because the precipitation step requires utilizing a ten-fold volume of DEE to extract the PCL-mal and peptide-PCL conjugate from the reaction solvent. In addition, the samples are centrifuged in 50 mL tubes, which further limits the reaction volume depending on the capacity of the available centrifuge. Dissolving PCL in NMP at 60 mg/mL is specific to this polymer in this solvent but can be adjusted to minimize volume for other polymers in other solvents.

11. As described in **Note 9**, the total reaction volume should be minimized. The exact volume necessary to fully dissolve the PMPI linker will vary depending on the reaction scale, but the minimum volume should be at least 1 mL to ensure full solubility and ability to accurately transfer the solution to the sealed reaction flask.

12. The supernatant will be a yellowish-orange color due to excess PMPI. PMPI does not precipitate as well as the PCL-mal in the DEE. However, minimizing the amount of time between precipitation and centrifugation steps is critical to avoid precipitation of unreacted PMPI.

13. The PCL-mal will have a slight yellowish color compared to the unmodified PCL, which is typically a white or off-white color.

14. PMPI proton peaks of should appear at higher ppm ranges (e.g., 6.5–8) without overlapping with the PCL peaks. ^{1}H NMR (CD_2Cl_2)—δ 7.6 (d, 2H, aromatic H orthogonal to maleimide), 7.3 (d, 2H, aromatic H orthogonal to carbamate), and 6.9 (s, 2H, maleimide vinyl). Additional information regarding peak assignments and integration can be found elsewhere [15].

15. The peptide can also be dissolved in DMSO to aid solubility depending on the peptide sequence. Negatively charged peptides tend to be less soluble in NMP and may dissolve better in DMSO. The peptide in DMSO can be added directly to the reaction.

16. The peptide-PCL conjugate may have a sticky or fibrous-like consistency. The precipitate can be manipulated with a spatula to break up the precipitate.

17. Depending on the solubility of the peptide in ultrapure water, you may need to add small amounts of acid or base to aid solubility depending on the number of charged residues and overall charge. The target pH for solubility of each peptide can be determined separately to avoid degrading the PCL, which hydrolyzes in acidic or basic conditions. If the precipitate is difficult to break down, sonication can be continued for longer times with heating to encourage peptide solubility.

18. The peptide-PCL conjugate may require sonication to fully dissolve in DMSO-d6 prior to NMR. The NMR spectra should be compared with NMR of the peptide alone and PCL-mal to verify and quantify conjugation. Notably, disappearance of the peak at 6.9 ppm corresponding to the maleimide vinyl protons suggests successful reaction with the thiol. Additional information regarding peak assignments and integration can be found elsewhere [15].

19. The peptide-PCL conjugate has been previously added at concentrations ranging between 1 and 12 mg/mL of the total electrospinning solution [15, 16]. The unmodified PCL concentration was not adjusted since the addition of the conjugate was typically negligible relative to the unmodified PCL concentration. Thus, the final concentration of total polymer and conjugate in solution may vary slightly depending on the concentrations used though no significant changes to bulk scaffold morphology were detected. High concentrations of peptide-PCL conjugate may interfere with the electrospinning process due to the peptide charge.

20. The viscosity of the polymer solution will be too high to pull through the needle. Remove the plunger, attach the capped needle, and slowly pour the solution directly into the syringe reservoir. Carefully reattach the plunger while inverting the syringe to avoid creating bubbles or expelling the solution from the needle end.

21. To minimize detachment of the tubing, keep one of the syringes immobilized on a syringe pump while manipulating the other. Care should be taken when pushing the solution through the tubing to prevent detachment and prevent bubble formation.

22. Minimal mixing will occur within the stopcock so solution 1 should be pushed through to ensure that the initial electrospinning solution only contains solution 1. The viscosity of the polymer solutions is high enough to prevent further mixing, especially while one solution is flowing.

23. Increasing or decreasing flow rates can affect fiber morphology and adequate fiber formation during electrospinning. Figure 3 illustrates a constant total flow rate since the increasing and

decreasing flow rates from each syringe are symmetric. The total flow rate does not have to be constant, but this will affect the fiber morphology and scaffold porosity.

Acknowledgments

This work was supported by start-up funds provided by Lehigh University. The author gratefully acknowledges helpful discussions about electrospinning with Dr. Geraldine Guex.

References

1. Langer R, Vacanti JP (1993) Tissue engineering. Science 260:920–926

2. James R, Toti US, Laurencin CT et al (2011) Electrospun nanofibrous scaffolds for engineering soft connective tissues. Methods Mol Biol 726:243–258

3. Castaño O, Eltohamy M, Kim H-W (2012) Electrospinning technology in tissue regeneration. Methods Mol Biol 811:127–140

4. Guex AG, Fortunato G, Hegemann D (2013) General protocol for the culture of cells on plasma-coated electrospun scaffolds. Methods Mol Biol 1058:119–131

5. Sill TJ, von Recum HA (2008) Electrospinning: applications in drug delivery and tissue engineering. Biomaterials 29:1989–2006

6. Liang D, Hsiao BS, Chu B (2007) Functional electrospun nanofibrous scaffolds for biomedical applications. Adv Drug Deliv Rev 59:1392–1412

7. Pham QP, Sharma U, Mikos AG (2006) Electrospinning of polymeric nanofibers for tissue engineering applications: a review. Tissue Eng 12:1197–1211

8. Wang X, Ding B, Li B (2013) Biomimetic electrospun nanofibrous structures for tissue engineering. Mater Today 16:229–241

9. Yoo HS, Kim TG, Park TG (2009) Surface-functionalized electrospun nanofibers for tissue engineering and drug delivery. Adv Drug Deliv Rev 61:1033–1042

10. Sun XY, Shankar R, Börner HG et al (2007) Field-driven biofunctionalization of polymer fiber surfaces during electrospinning. Adv Mater 19:87–91

11. Gentsch R, Pippig F, Schmidt S et al (2011) Single-step electrospinning to bioactive polymer nanofibers. Macromolecules 44:453–461

12. Lancuški A, Bossard F, Fort S (2013) Carbohydrate-decorated PCL fibers for specific protein adhesion. Biomacromolecules 14:1877–1884

13. Lancuški A, Fort S, Bossard F (2012) Electrospun Azido-PCL Nanofibers for enhanced surface functionalization by click chemistry. ACS Appl Mater Interfaces 4:6499–6504

14. Sundararaghavan HG, Burdick JA (2011) Gradients with depth in electrospun fibrous scaffolds for directed cell behavior. Biomacromolecules 12:2344–2350

15. Chow LW, Armgarth A, St-Pierre J-P et al (2014) Peptide-directed spatial organization of biomolecules in dynamic gradient scaffolds. Adv Healthc Mater 3:1381–1386

16. Harrison RH, Steele JAM, Chapman R et al (2015) Modular and versatile spatial functionalization of tissue engineering scaffolds through fiber-initiated controlled radical polymerization. Adv Funct Mater 25:5748–5757

17. Wade RJ, Bassin EJ, Gramlich WM et al (2015) Nanofibrous hydrogels with spatially patterned biochemical signals to control cell behavior. Adv Mater 27:1356–1362

18. Sant S, Hancock MJ, Donnelly JP et al (2010) Biomimetic gradient hydrogels for tissue engineering. Can J Chem Eng 88:899–911

19. Boyer C, Granville A, Davis TP et al (2009) Modification of RAFT-polymers via thiol-ene reactions: a general route to functional polymers and new architectures. J Polym Sci A Polym Chem 47:3773–3794

Chapter 4

Low-Temperature Deposition Modeling of β-TCP Scaffolds with Controlled Bimodal Porosity

E. Papastavrou, P. Breedon, and D. Fairhurst

Abstract

Low-temperature deposition modeling (LDM), otherwise termed freeze-form extrusion fabrication or rapid freeze prototyping, involves dispensing an aqueous-based ceramic paste or polymeric hydrogel along predefined paths in subzero ambient temperatures, followed by freeze-drying. The solidification of the material after the deposition of each layer enables large parts to be built without the need for organic binders, which can often have cytotoxic effects. Freeze-dried parts obtained from LDM typically exhibit pores with openings that range in average between 1 and 40 μm. The technique offers the ability to control their size distribution and orientation through varying a number of processing and material parameters. Herein, we describe the construction of an LDM system from readily available electromechanical components, as well as the preparation of a β-TCP paste formulation with the appropriate flow characteristics for fabricating hierarchical scaffolds with tailorable bimodal porosity for applications in bone tissue engineering.

Key words Low-temperature deposition modeling, Additive manufacturing, Ice-templating, Hierarchical scaffolds, Hard tissue engineering, Bio-ceramics, Bimodal porosity, Freeze-drying, Thixotropic

1 Introduction

Among the vast range of additive manufacturing techniques available for fabricating hard tissue engineering constructs, filament-based techniques constitute the most versatile and economic method. They can be easily combined with conventional scaffold manufacturing techniques, opening up countless possibilities in structuring biomaterials for applications in bone tissue regeneration. Hierarchical porous scaffolds have demonstrated superior bone forming activity in vitro [1]. A highly interconnected pore network allows for efficient nutrient delivery and cell migration. Ideally, a scaffold should exhibit hierarchical porosity of 60–65%, with approximately 60% of its pores ranging from 150 to 400 μm and at least 20% smaller than 20 μm [2, 3]. Microporosity is desirable because it offers a larger specific surface area, accelerating ion exchange and biological apatite deposition and increasing the

Kanika Chawla (ed.), *Biomaterials for Tissue Engineering: Methods and Protocols*, Methods in Molecular Biology, vol. 1758, https://doi.org/10.1007/978-1-4939-7741-3_4, © Springer Science+Business Media, LLC, part of Springer Nature 2018

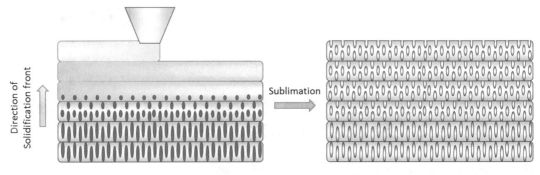

Fig. 1 Visual representation of the LDM process

number of adsorption sites for bone-inducing proteins [4]. It has also been demonstrated that the presence of interconnected micropores (0.5–10 μm diameter) promotes osseointegration at different scales, as it assists osteogenic cells in populating the internal surfaces of a scaffold [5]. Pore size can also have an effect on the osteogenic path, favoring direct osteogenesis in large pores and osteochondral ossification in small pores [6].

In low-temperature deposition modeling, the shape and internal architecture of a scaffold are controlled through a freeform fabrication process (robocasting), while the desirable microstructural characteristics, such as surface roughness and porosity, are achieved through freezing (ice templating). A robotic dispensing system deposits an aqueous bio-ceramic paste in an ambient temperature that is below the freezing point of water. As the paste freezes, its particles are expelled from the moving solidification (freezing) front and accumulate between the growing ice crystals. The frozen part is lyophilized (freeze-dried) to sublimate the freezing medium, leaving behind a highly porous structure (see Fig. 1). Finally, depending on the application, the dried green body can be sintered to consolidate its particles.

As with ice templating, a large range of microstructure formations can emerge by controlling the following parameters:

- The freezing-front velocity is the most important operational condition. It is determined by the cooling rate of the sample. It has been observed that lamellar wall spacing and average pore size increase with decreasing cooling rate [7].

- Addition of dispersants, binders, pH, and viscosity of the slurry [8].

- Thermal conductivity of constituents.

- The presence of additives such as gelatine, glycerol, and sucrose can alter the shape, size, and roughness of micropores [9].

- Concentration (solid loading) and size of particles: High solid loading decreases the global porosity of the sample, while finer particles can replicate the most intricate features of ice crystals [10].

This particular feature together with the possibility of preserving bioactive substances, since no heating or melting is required, makes LDM highly suitable for tissue engineering applications. Xiong et al. [11] have successfully demonstrated this by fabricating TCP/poly(L-lactic) acid (PLLA) scaffolds of 89.6% average porosity. The scaffolds, besides their interconnected macroporosity, also exhibited micropores with an average size of 5 μm. Dorj et al. [12] employed LDM to produce chitosan/nano-bioactive glass scaffolds with dual-pore structure, which displayed satisfactory cell adherence and proliferation. Pham et al. [13] and Lim et al. [14] have reported on the effect of freezing rate on the size and orientation of micropores within scaffolds made from chitosan. It was observed that low freezing rates resulted into large pores that were perpendicular to the direction of the dispensed path, orientated along temperature gradients.

As a model material, we have developed a highly thixotropic β-TCP paste formulation with low solid loading that demonstrates good injectability and shape retention after deposition. Its viscoelastic properties can be tuned through the weight ratio of its constituent hydrocolloids in the biphasic mixture, namely sodium alginate and xanthan gum. This constitutes a versatile way of tailoring the rheological properties of a wide range of ceramic biomaterials in a single step.

A commercial fused deposition modeling (FDM) 3D printer has been modified to accommodate the extrusion of colloidal β-TCP pastes in subzero temperatures. The modification involved replacing the polymer extruder with a pressure-driven dispensing system to produce a simple and economical design that offers high precision and repeatability. A schematic diagram of the printhead is illustrated in Fig. 2. This chapter provides a step-by-step guide for the assembly of a low-temperature deposition modeling system and additional basic operating instructions for fabricating synthetic hard tissue constructs with controlled bimodal porosity.

2 Materials

2.1 Bio-ceramic Paste Formulation

2.1.1 All Chemicals and Reagents Used as Received

1. β-TCP powder (Sigma Aldrich, particle diameter 1–5 μm): Other bio-ceramic materials can also be used, such as HA and Bioglass.

2. Sodium alginate (SA) (food grade).

3. Xanthan gum (XG) (food grade).

4. Distilled water.

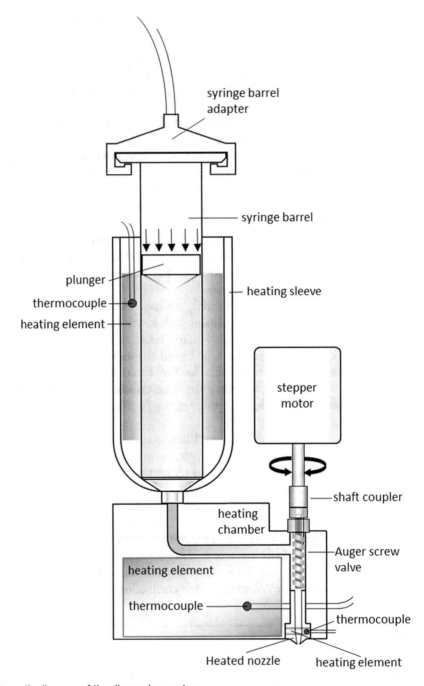

Fig. 2 Schematic diagram of the dispensing system

2.2 Equipment

2.2.1 Robocasting System

1. Laser cutter.
2. Acrylic sheet 220×400 mm.
3. Cyanoacrylate adhesive.
4. 3 mm thick sheet of PTFE (40×112.5 mm).

5. Large polypropylene syringe barrel (960 cm³) (40 mm internal diameter).

6. Nitrile rubber foam insulation (tubular and sheet with adhesive).

7. Glass wool.

8. Two 100×75 mm silicone heater mats (12 V, 7.5 W) with 12 V power supply.

9. K-type thermocouples.

10. Two digital temperature controllers.

11. NEMA 17 stepper motor.

12. Shaft coupler (5 mm circular to 5 mm square shaft).

13. Delrin Disposable Material Path (Techcon TS5000DMP).

14. Standard Rep Rap "J-Head MkV" hot end (0.5 mm), typically used in fused deposition modeling systems.

15. Kapton® tape.

16. PTFE tape.

17. Female luer-lock coupler.

18. Male luer lock-to-female 5/16″-24 adapter.

19. Compressor or pressurized air supply with pressure regulator.

20. Disposable 55 cm³ syringe barrels.

21. Syringe barrel adapter with airline hose.

22. Disposable cartridge plungers.

23. Motorized xyz gantry ($275 \times 275 \times 210$ mm).

24. Silicone thermal pad with heat transfer coefficient of 2.2 W/(m^2 K).

25. Control unit: Arduino Mega 2560 with RAMPS 1.4 shield.

26. 12 and 18 mm M3 screws and nuts.

27. Grasshopper plug-in for Rhinoceros®, Robert McNeel & Associates.

2.2.2 Additional Equipment

1. Industrial freezer or preferably environmental chamber that can reach temperatures down to −20 °C.

2. LED light source and web camera (optional, in case the environmental chamber does not feature a viewing window).

3. Freeze-dryer (−20 °C condenser temperature, minimum achievable vapor pressure of 0.021 mbar).

4. Kiln (optional).

2.2.3 Paste Preparation

1. Mortar and pestle.

2. Stainless steel 62 μm sieve.

3. Rubber kidney.

4. Small plastering trowel or spatula.

3 Methods

3.1 Low-Temperature Deposition Modeling System Assembly

1. Laser-cut the acrylic components. Technical drawings are provided in Figs. 3, 4, and 5. Adapt the design accordingly, so that it fits individual printhead mounts.

2. Glue the components together with cyanoacrylate adhesive. These are divided into two groups, indicated with a letter (A or B), followed by a number that represents their order in the stack (see Figs. 5 and 6a). Special attention should be paid to their alignment.

3. Place the coupler over the stepper motor's shaft.

4. Mount the stepper motor on part B with four 18 mm long M3 screws.

5. Join parts A and B with M3 screws and nuts. Proceed with the rest of the assembly (Fig. 6b).

6. Press the rotor (auger screw) of a disposable material path onto the square end of the shaft coupler. Place the stator over it and slot into the bore of part A, pushing against the motor shaft.

7. For the heated nozzle assembly, insert and secure the thermocouple and heating element into the 0.5 mm J-Head MkV hot end with Kapton® tape. Screw the female 5/16″-24-to-male luer fitting on its threaded top end. Seal the thread with PTFE tape to prevent leakage (Fig. 6d).

8. Connect the heated nozzle to the stator with a female luer-lock coupler.

9. Affix nitrile rubber foam (NRF) insulation on the acrylic parts that make up the heating chamber for the feed passage and valve (see Fig. 4 for technical drawing). Attach a K-type thermocouple on a 100×75 mm silicone heater mat (12 V, 7.5 W) and position both inside the chamber (Fig. 6c). Reinforce the thermal insulation with glass wool where necessary. Seal the casing's bottom face with a 3 mm thick piece of PTFE (Fig. 5). Counterbore the holes near the edges so that the screws do not protrude beyond the unit's bottom surface (Fig. 6f). Adjust the height of the stepper motor's shaft coupler accordingly.

10. The heating sleeve for the pressurized paste reservoir is constructed as follows: cut the threaded tip off a 960 cm³ polypropylene syringe barrel and reduce its height to 15 cm. Wrap it in 6 mm of NRF insulation with self-adhesive backing. Attach a K-type thermocouple on a 100×75 mm silicone heater mat (12 V, 7.5 W) and position both inside the modified barrel.

11. Mount the dispensing system on the motorized xyz gantry.

12. Position the entire unit inside the industrial freezer, leaving outside its control unit, temperature controllers, and power supplies.

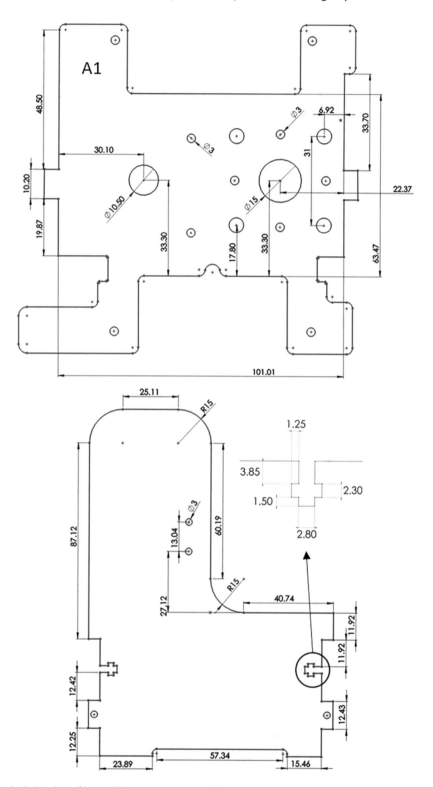

Fig. 3 Technical drawing of base (A1) and rear parts of the dispensing system

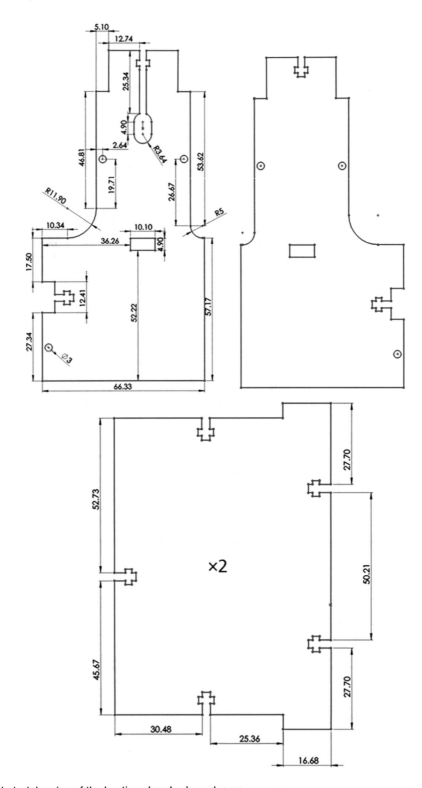

Fig. 4 Technical drawing of the heating chamber's enclosure

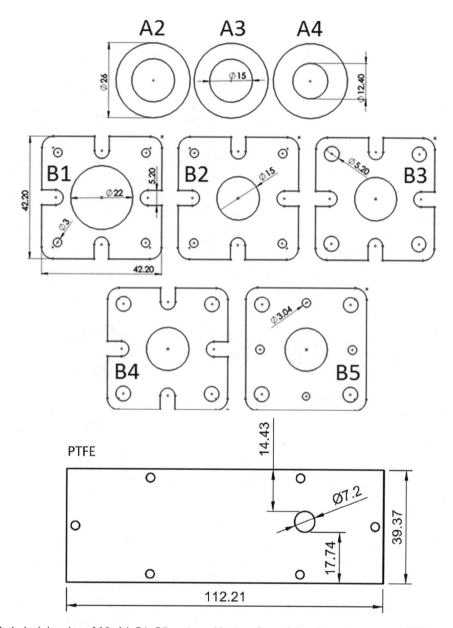

Fig. 5 Technical drawing of A2–A4, B1–B5 parts and bottom face of the dispensing system (PTFE)

13. Wire the stepper motors, end stops, and hot end to the Arduino-based control unit. Use wiring of sufficient length and advance through the drain hole at the bottom of the freezer. Connect the silicone heater mats and thermistors to their respective temperature controllers and power supplies. Keep cables away from moving parts.

14. Position the web camera and LED light source in the freezer, ensuring an optimal view of the print area (Fig. 6g).

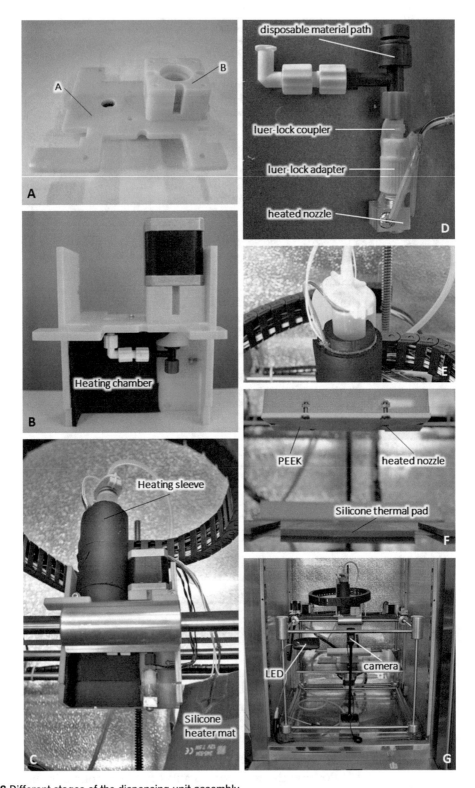

Fig. 6 Different stages of the dispensing unit assembly

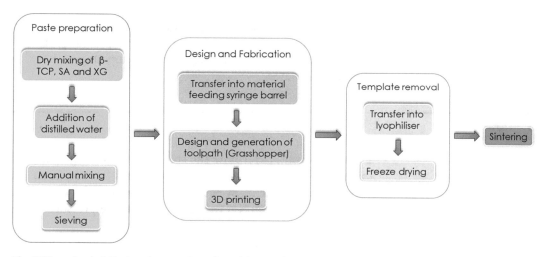

Fig. 7 Flow chart of the low-temperature deposition modeling process

15. Place a silicone thermal pad with heat transfer coefficient of 2.2 W/(m² K) on the print bed. This is the substrate on which the part is 3D printed (Fig. 6f).

3.2 Operating Instructions for the Low-Temperature Deposition Modeling System

1. The flow diagram in Fig. 7 illustrates the key stages of the LDM procedure. The system must be calibrated before operation (*see* **Notes 1–4**).

2. Generate the tool path (*see* **Note 5**).

3. Turn on the air supply and adjust the pressure.

4. Actuate the extruder's motor until the paste reaches the tip of the heated nozzle. Apply some petroleum jelly to prevent the paste from drying and blocking its orifice. The air supply can be switched off until the system reaches the set temperature.

5. Switch on the freezer/environmental chamber and set to required temperature (*see* **Note 6**).

6. Turn on the power supplies and digital temperature controllers corresponding to the heating sleeve and chamber (*see* **Note 7**).

7. Switch on the heated nozzle and maintain its temperature at 20 °C until the freezer reaches the set value.

8. 3D printing should commence after the system has reached equilibrium (*see* **Notes 8–11**).

9. Transfer the part, together with the substrate (thermal pad), into the lyophilizer's condenser. It is important that this transfer be done rapidly, to minimize the risk of annealing or melting of the ice matrix. The duration of the freeze-drying process is dictated by the size of the specimen.

10. Sinter the part, if necessary, to consolidate its particles using the appropriate temperature program. The β-TCP paste formulation described herein requires sintering due to its high solubility in water.

4 Notes

1. For each motor, determine the steps corresponding to a fixed distance along the x-, y-, and z-axes. Calculate the speed at which the extruder must rotate in relation to the motors that control its movement on the XY plane and the pressure inside its feeding section.

2. Setting the pressure inside the syringe barrel too low can significantly limit the attainable range of printing speeds. On the other hand, increasing it beyond a certain level can result into a continuous flow of paste (drooling) after the rotor becomes stationary. This problem can be addressed through decompressing the nozzle's chamber and introducing a printhead cleaning stage between dispensing cycles. Decompression (suck back) is achieved through reversing the direction of rotation.

3. Excessive "drooling" can also be a sign of the auger screw requiring replacement, as it is progressively worn down by the ceramic particles present in the paste.

4. To minimize the extrusion's pulsation, correct the air pressure and/or the motor's RPM for a given printing speed.

5. Our study employed parametric design software for generating tool paths (Grasshopper plug-in for Rhinoceros®, Robert McNeel & Associates). This plug-in provides a visual programming environment and most importantly the ability to view in real time the resulting geometry during the creation or modification of a script. Additionally, it enables designers to define parametric relationships between different elements of a model.

6. High freezing temperatures result in parts containing large lamellar pores, whereas low freezing temperatures produce small cellular pores, as can be seen in Fig. 8.

7. To prevent the paste from freezing, the set-point value of the silicone heater mats should be varied depending on ambient temperature. These values can be determined by measuring the actual temperature inside the heating chamber and sleeve once the system has reached equilibrium.

8. A temperature gradient exists along the heated nozzle, whose lower limit is dictated by the temperature inside the freezer.

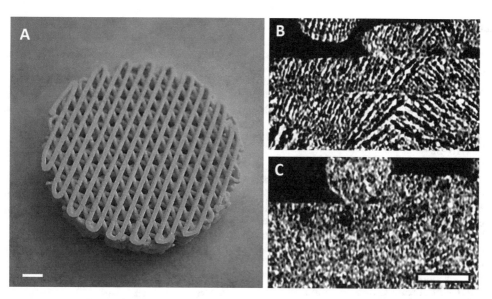

Fig. 8 (**a**) 3D lattice fabricated with LDM (scale bar = 2 mm), μCT vertical cross sections illustrating the lamellar microstructure obtained at (**b**) 5 °C compared to (**c**) the cellular microstructure obtained at 20 °C. Scale bar = 500 μm (μCT images provided by Dr. Nikolaus Nestle, BASF)

Consequently, low extrusion rates can shift the extrudate's temperature toward this lower limit. At high printing speeds, the temperature of the material is expected to be closer to that which the nozzle is set to. During LDM, it is advisable that the temperature is maintained as low as possible. High temperatures can have a detrimental effect on the overall quality of the final part.

9. A short pause after the deposition of each layer is essential for the successful fabrication of a part, as this allows for its solidification and cooling before the addition of another layer.

10. Set the layer height to approximately 75% of the nozzle's nominal diameter. Layer thickness can be affected by ambient temperature. A possible reason is the contraction of the setup's mechanical components and more specifically the lead screw that actuates the print bed along the z-axis. Avoid using power transmission components with polymeric parts, as they contract more than metals.

11. A certain amount of paste can sometimes accumulate between the stepper motor and the DMP. It is recommended to position a rubber seal over the shaft between parts B2 and B4. This is not featured in the present design.

References

1. Yun H, Kim S, Hyeon Y (2007) Design and preparation of bioactive glasses with hierarchical pore networks. Chem Commun (Camb) 21:2139–2141

2. Nandi SK, Roy S, Mukherjee P et al (2010) Orthopaedic applications of bone graft & graft substitutes: a review. Indian J Med Res 132:15–30

3. Dorozhkin SV (2013) Calcium orthophosphate-based bioceramics. Materials (Basel) 6:3840–3942

4. Mastrogiacomo M, Scaglione S, Martinetti R et al (2006) Role of scaffold internal structure on in vivo bone formation in macroporous calcium phosphate bioceramics. Biomaterials 27:3230–3237

5. Lan Levengood SK, Polak SJ, Wheeler MB et al (2010) Multiscale osteointegration as a new paradigm for the design of calcium phosphate scaffolds for bone regeneration. Biomaterials 31:3552–3563

6. Hannink G, Arts JJC (2011) Bioresorbability, porosity and mechanical strength of bone substitutes: what is optimal for bone regeneration? Injury 42(Suppl 2):S22–S25

7. Farhangdoust S, Zamanian A, Yasaei M, Khorami M (2013) The effect of processing parameters and solid concentration on the mechanical and microstructural properties of freeze-casted macroporous hydroxyapatite scaffolds. Mater Sci Eng C 33:453–460

8. Zhang Y, Zhou K, Bao Y, Zhang D (2013) Effects of rheological properties on ice-templated porous hydroxyapatite ceramics. Mater Sci Eng C 33:340–346

9. Deville S (2013) Ice templating, freeze casting: beyond materials processing. J Mater Res 28(17):2202–2219

10. Wegst UGK, Schecter M, Donius AE, Hunger PM (2010) Biomaterials by freeze casting. Philos Trans A Math Phys Eng Sci 368:2099–2121

11. Xiong Z, Yan Y, Wang S et al (2002) Fabrication of porous scaffolds for bone tissue engineering via low-temperature deposition. Scr Mater 46:771–776

12. Dorj B, Park J-H, Kim H-W (2012) Robocasting chitosan/nanobioactive glass dual-pore structured scaffolds for bone engineering. Mater Lett 73:119–122

13. Pham CB, Leong KF, Lim TC, Chian KS (2008) Rapid freeze prototyping technique in bio-plotters for tissue scaffold fabrication. Rapid Prototyp J 14:246–253

14. Lim TC, Bang CP, Chian KS, Leong KF (2008) Development of cryogenic prototyping for tissue engineering. Virtual Phys Prototyp 3:25–31

Chapter 5

Three-Dimensional Hydrogel-Based Culture to Study the Effects of Toxicants on Ovarian Follicles

Hong Zhou and Ariella Shikanov

Abstract

Various toxicants, such as drugs and their metabolites, can cause potential ovarian toxicity. As the functional units of the ovary, ovarian follicles are susceptible to this type of damage at all developmental stages. Studying the effects of toxicants on ovarian follicles is an important task. Three-dimensional (3D) hydrogels, such as fibrin alginate interpenetrating networks (FA-IPNs), can support ovarian follicle culture in vitro for extended periods of time and serve as a suitable tool for studying ovotoxicity. Growing follicles encapsulated in the FA-IPN can proteolytically degrade the fibrin component in the FA-IPN. The degradation of fibrin mirrors the follicle growth and serves as a surrogate reporter for follicle health. The speed of fibrin degradation can be further controlled by aprotinin, a small molecule that inhibits plasmin-driven proteolytic degradation, which further expands the application of the described system. In this chapter, we describe methods to (1) isolate and encapsulate mouse ovarian follicles in FA-IPN, (2) follow follicle growth and development in vitro, and (3) evaluate the effects of toxicants on folliculogenesis using fibrin degradation.

Key words Ovarian follicles, Fibrin, Alginate, Aprotinin, Toxicology

1 Introduction

Mammalian ovaries contain a finite number of follicles. As the functional units of the ovaries, ovarian follicles are responsible for a female's fertility and ovarian endocrine functions. In vitro follicle growth (IVFG) is an experimental tool that recapitulates key events of mammalian oogenesis and folliculogenesis in vivo. By monitoring follicle survival and growth, antral cavity formation, steroidogenesis, and ability of producing a mature egg, IVFG can serve as an accurate bioassay for studying essential biological processes related to female reproductive functions [1–5].

Various IVFG culture systems have been developed aiming to grow follicles from an immature stage to fully mature and fertilizable oocytes, which have been successful in mouse, rat, bovine, goat, nonhuman primates, and human [6–14]. Three-dimensional

Kanika Chawla (ed.), *Biomaterials for Tissue Engineering: Methods and Protocols*, Methods in Molecular Biology, vol. 1758, https://doi.org/10.1007/978-1-4939-7741-3_5, © Springer Science+Business Media, LLC, part of Springer Nature 2018

IVFG systems in particular provide physical support to preserve the architecture of developing follicles and maintain oocyte-somatic cell connections, thus promoting survival of early-stage follicles [15]. Encapsulated in either natural biomaterials such as alginate [16, 17] and fibrin-alginate [18, 19] or synthetic macromers such as polyethylene glycol (PEG) [20], ovarian follicles are cultured in invivo-like conditions, allowing for continuous morphological, genetic, and biochemical sampling.

An important application of IVFG is serving as a bioassay for reproductive toxicology screening. Safety information on new chemicals that are introduced into the US market each year is lacking due to the cost and time required for animal testing. Performing a complete set of regulatory tests for a single chemical requires thousands of animals and costs millions of dollars [1]. IVFGs have been employed to determine how follicles are affected by drugs/their metabolites, and environmental factors, such as doxorubicin (DXR), organochloride pesticides (methoxychlor (MXC)), and environmental pollutants (7,12-dimethylbenz[a]anthracene (DMBA)). Compared to animal studies, IVFG represents a more simple and rapid tool that can be used to screen the effects of potential ovotoxic compounds on reproductive function and health [1, 4, 21], which can be useful to provide guidance to regulatory agencies as well as expectant mothers.

In this chapter, we describe a fibrin-alginate-based hydrogel system [18] that has been developed for IVFG. In the fibrin alginate interpenetrating network (FA-IPN) system, the slowly degrading alginate is biologically inert to ovarian tissue, thus providing structural support for the follicle. Ovarian follicles encapsulated in FA-IPN secrete plasminogen activators [22, 23], and can therefore degrade the fibrin component of the FA-IPN. The semi-degradable FA-IPNs provide a dynamic mechanical environment that facilitates follicle outward expansion while maintaining follicle architecture. Such systems have also been proven to enable IVFG in larger species such as baboon [24] and rhesus macaque (*Macaca mulatta*) [25]. The degraded fibrin appears as an optically clear ring around the encapsulated follicle as the follicle culture progresses [18, 19, 26], correlating with the health and survival of the encapsulated follicles. Furthermore, a calibrated addition of aprotinin, which is a plasmin inhibitor, to the culture media results in controlled fibrin degradation and therefore broadens the applications for this system [26] to screen for potential ovotoxicity induced by drugs and/ or their metabolites and environmental toxicants.

2 Materials

2.1 General Materials

1. One pair of straight fine scissors, 26 mm.

2. One pair of straight fine scissors, 24 mm.

3. One pair of straight forceps, #5.

4. One pair of curved forceps, #7.

5. Dissecting microscope with a heating stage.

6. Inverted imaging microscope with imaging software such as ImageJ.

2.2 Ovarian Follicle Isolation

1. First-generation female hybrid offsprings of two inbred strains: C57BL/6JRccHsd (maternal) CBA/JCrHsd (paternal), 12–14 days of age.

2. 70% Ethanol and dissecting mats.

3. Dissection media (DM): Leibovitz's L-15 medium, heat-inactivated fetal bovine serum (FBS), and Pen/Strep.

4. Maintenance media (MM): Minimum essential media α (αMEM), FBS, and Pen/Strep.

5. Two 35 × 10 mm sterile petri dishes.

6. One 60 × 10 mm sterile petri dishes.

7. Center Well Dishes for IVF: Use one (1) dish per each ovary.

8. Sterile 1.5 mL microcentrifuge tubes.

9. Two sterile insulin syringes with 27½ G needles.

2.3 Ovarian Follicle Encapsulation

1. Dulbecco's phosphate-buffered saline 1× (DPBS, no calcium, no magnesium).

2. Sterile alginate aliquots: Nalgin MV-120 alginate, activated charcoal.

3. Sterile fibrinogen and thrombin aliquots.

4. Tris-buffered saline (TBS) 25 mM with 0.15 M NaCl: Tris–HCl, Tris base, and NaCl.

5. TBS 25 mM with 50 mM $CaCl_2$: Tris–HCl, Tris base, $CaCl_2 \bullet 2H_2O$, and NaCl (cross-linking solution).

6. Baxter Tisseel Fibrin Sealant: Sealer Protein Solution (fibrinogen, total protein: 96–125 mg/mL, with synthetic aprotinin: 2250–3750 KIU/mL) and thrombin solution (human thrombin: 400–625 IU/mL with $CaCl_2$: 36–44 μmol/mL).

7. Sterile 1.5 mL microcentrifuge tubes.

8. Center Well Dishes for IVF dishes.

9. Steriflip conical tubes.

10. Sterile cellulose membrane syringe filters.

2.4 Ovarian Follicle Culture

1. DPBS.

2. Growth media (GM): αMEM, 1 mg/mL fetuin, 3 mg/mL bovine serum albumin (BSA), 5 μg/mL insulin, 5 μg/mL

transferrin, and 5 ng/mL selenium, and 10 mIU/mL recombinant human follicle-stimulating hormone (rhFSH).

3. Sterile aprotinin aliquot: Aprotinin, $1\times$ DPBS.

4. Sterile 1.5 mL microcentrifuge tubes.

3   Methods

All dissections should be performed in L15-based media (buffered for ambient air), on a 37 °C (temperature control) heated stage, and on a clean bench (laminar flow hood) to minimize potential contamination. Follicle isolation should take no longer than 30 min per ovary. For optimal results, keep the resected tissue outside the incubator for less than 1 h. To minimize pH changes, limit exposure of αMEM-based media to ambient air. All animals are treated in accordance with the guidelines and regulations set forth by the national and institutional IACUC protocols.

3.1   Ovarian Follicle Isolation

1. Prior to the beginning of the experiment, sterilize all the equipment by spraying 70% ethanol and air-drying on the clean bench or in the laminar hood.

2. Prepare DM by supplementing L-15 with 1% (v/v) FBS and 0.5% (v/v) Pen/Strep. Gently invert to mix well (*see* **Note 1**). Warm up the DM to 37 °C in a water bath (*see* **Note 2**).

3. Prepare MM by supplementing αMEM with 1% (v/v) FBS and 0.5% (v/v) Pen/Strep. Gently invert to mix well (*see* **Note 3**). Prepare IVF dishes with 1 mL of MM in the center ring and 3 mL of MM in the outer ring. Utilize one IVF dish for each isolated ovary. In addition, prepare another 35 mm dish with 1 mL of MM. Pre-equilibrate all the dishes for 20 min in the incubator prior to the beginning of the isolation.

4. Euthanize one female (12–14 days of age), according to the institutional IACUC-approved protocols. Remove both ovaries from the animal. In order to assure minimal damage to the ovaries, remove parts of the oviduct and uterus around the ovary as well. Place roughly dissected ovaries into a 35 mm dish with DM.

5. Use a dissecting scope and insulin syringes to separate the ovaries from uterus, fat pad, and bursa (Fig. 1a) by placing one needle at the intersection of the bursa and the oviduct to anchor the reproductive tract in place, and dissecting with the other needle. Place the second needle directly next to the first but only grip the thin membrane of the bursa. Carefully nick the bursa to expose the entire ovary. Transfer clean ovaries using curved forceps or a pipette with a blunt tip (Fig. 1b) into the 35 mm dish with pre-equilibrated MM. If using forceps,

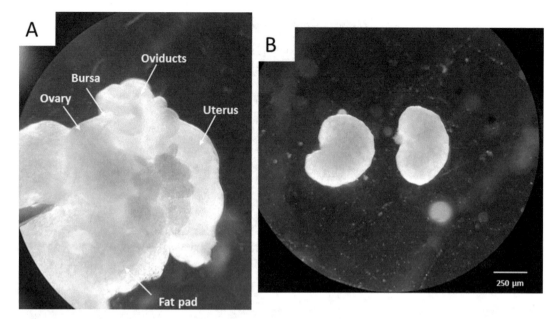

Fig. 1 Isolating ovaries from surrounding tissues. (**a**) Resected tissues with one ovary embedded. (**b**) Isolated ovaries clean from tissues

try to not squish or damage the ovaries by applying excessive force. Repeat this process for both ovaries. Transfer one ovary to a 35 mm dish containing warm MM and place it in the incubator. Keep the other ovary in DM and start follicle isolation [17].

6. Set a timer for 30 min. Start isolating follicles by using two insulin syringes with 27½ G needles. With one syringe in your nondominant hand, anchor the ovary to the bottom of the dish. And with a syringe in your dominant hand, gently tease and "flick" individual follicles from the ovary (Fig. 2a). Try to remove most of the surrounding ECM without puncturing the follicle. Dissect out as many follicles as possible within the 30 min. Transfer intact isolated secondary follicles (120–135 μm in diameters, 2–3 layers of granulosa cells) to the outer ring of an IVF dish (Fig. 2b, c) with MM. When transferring, use a P10 pipette to carefully pick up isolated follicles with minimum media. To avoid follicles sticking to the walls of the same pipette tip and/or to each other, aspirate a little media first to the very end of the tip and while aspirating the follicles alternate between media and each follicle. When expelling the follicles into the outer ring of the IVF dish, use a sweeping motion such that the follicles in the tip will end up at different spots to avoid sticking to each other (*see* **Note 4**).

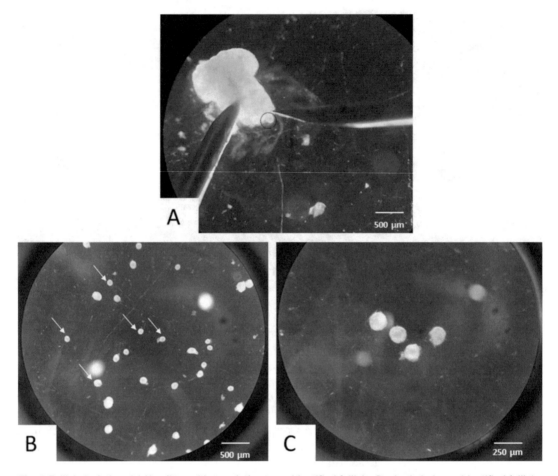

Fig. 2 Follicle isolation. (**a**) Needle positions relative to an identified follicle. Dashed circle: an identified follicle. The left needle is used to anchor the tissue to the bottom of the plate while the right needle is used to tease and "flick" the follicle away from the ovary. (**b**) Isolated follicles under 50×. Try to keep isolated follicles apart to prevent them from sticking back together. White arrows: example follicles. (**c**) Suitable follicles examined under 100× for proper diameters and intact morphology

7. Repeat Subheading 3.1, **steps 1–6**, for the remaining ovaries to complete the isolation.

3.2 Ovarian Follicle Encapsulation (See Note 5)

1. The alginate component of the FA-IPN is Nalgin MV-120 sodium alginate. To prepare sterile alginate aliquots (*see* **Note 6**), dissolve alginate in deionized water at a concentration of 0.1% (w/v) (*see* **Note 7**). Add activated charcoal [0.5 g charcoal/(g alginate)] to the alginate solution to remove organic impurities. Stir the mixture at room temperature for 15 min and let settle overnight. Following the charcoal treatment, sterile filter alginate solution through 0.20 μm fast vacuum filtration units. Transfer the sterile alginate solution into Steriflip conical tubes to lyophilize. Aliquot the lyophilized

sterile alginate into sterile 1.5 mL microcentrifuge tubes. One day before follicle isolation, reconstitute the sterile alginate aliquot with 1× DPBS to achieve 0.8% (w/v). Leave in the laminar hood overnight to completely dissolve (*see* **Note 8**). On the day of experiment, vortex the alginate solution for 30 s to thoroughly mix. Briefly centrifuge it afterwards to prevent microbubbles.

2. Prepare 25 mM TBS with 0.15 M NaCl by dissolving 3.5618 g of Tris–HCl, 0.2786 g of Tris base, and 8.7750 g of NaCl in 1 L of deionized water. Titrate pH to 7.4 at room temperature. Sterile filter a proper amount of the solution before use.

3. Prepare the fibrinogen according to kit instructions (*see* **Note 9**). First, reconstitute fibrinogen to 100 mg/mL (total protein) with the provided fibrinolysis inhibitor solution containing 3000 KIU/mL aprotinin. Then, dilute the reconstituted fibrinogen down to 50 mg/mL with additional 25 mM TBS with 0.15 M NaCl. Aliquot the diluted fibrinogen to sterile 1.5 mL microcentrifuge tubes as 100 µL/tube. Reconstitute thrombin to 500 IU/mL in the 40 µmol/mL $CaCl_2$ solution provided in the kit. Aliquot 50 µL/tube thrombin to sterile 1.5 mL microcentrifuge tubes (*see* **Note 10**).

4. Prepare 25 mM Tris-buffered saline (TBS) with 50 mM $CaCl_2$ as the cross-linking solution. To make 1 L of this solution, dissolve in the deionized water the following: 3.5618 g of Tris–HCl, 0.2786 g of Tris base, 7.3505 g of $CaCl_2 \bullet 2H_2O$ (adjust this amount if using anhydrous $CaCl_2$), and 2.9220 g of NaCl (*see* **Note 11**). Titrate pH to 7.4 at room temperature. Sterile filter the solution prior to use.

5. After all the isolated follicles are transferred to MM (*see* **Note 12**), thaw fibrinogen and thrombin aliquot(s) based on the number of follicles isolated. To prepare fibrinogen/alginate solution (FA) for encapsulation, mix fibrinogen (50 mg/mL), 0.8% alginate MV (vortex prior to use), and 1× DPBS at 1:1:2 ratio (*see* **Note 13**). Pipette up and down to thoroughly mix (*see* **Note 14**). It is normal if the mixed solution appears milky. Dilute thrombin stock solution with sterile solution of 25 mM TBS with 50 mM $CaCl_2$ to reach a final concentration of thrombin at 25 mg/mL. Transfer this working thrombin solution to the inner ring of a new IVF dish. In the outer ring, put a droplet of approximately 10–20 µL of FA as the washing droplet. Prepare another 100 µL of FA droplet at a different spot in the outer ring of the same IVF dish as the encapsulating droplet (Fig. 3) (*see* **Note 15**).

6. Inspect the follicles meant for encapsulation under 50×–80× magnification. Use the following criteria to identify healthy follicles: (1) 2–3 layers of granulosa cells; (2) in the right size

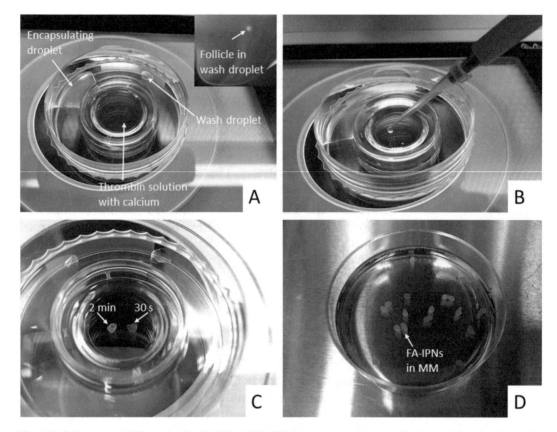

Fig. 3 Follicle encapsulation setup for FA-IPN. (**a**) FA-IPN encapsulation setup with the washing droplet on the right and the encapsulating drop on the left. (**b**) Dropping the FA-IPN with a follicle into the TBS solution with calcium. (**c**) FA-IPN beads in the cross-linking solution the cross-linking progresses. (**d**) FA-IPN beads in MM

range; (3) with no separation between the oocyte and granulosa cells; (4) have round oocytes. Some theca cells or extra-cellular matrix are fine (*see* **Note 16**).

7. Using a dissecting scope, carefully pick up a good follicle using a P10 pipette with minimum media and transfer this follicle to the wash droplet (Fig. 3a). Gently pipette it up and down to remove excess media from the follicle before transferring it to the encapsulating droplet (*see* **Notes 17** and **18**). Work quickly because the FA solution is not viscous enough to prevent follicles settling to the bottom of the dish. If the follicle falls through the droplet of FA solution, it will be difficult to pick it back up and encapsulate because follicles tend to stick to the plastic.

8. After transferring the follicle to the encapsulating droplet, use a P10 pipette set to 7.5 µL. First, gently fill about half of the tip (~3.5 µL) with FA. Then, aspirate the follicle and complete the rest of the 7.5 µL with more FA.

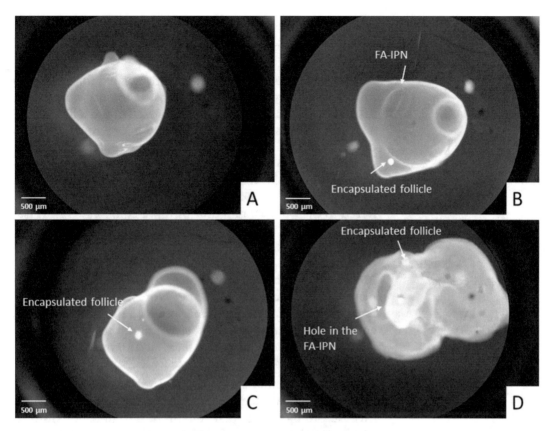

Fig. 4 Follicles encapsulated in FA-IPNs. (**a**) Empty bead. (**b**) Follicle encapsulated at the edge of the bead. (**c**) Follicle encapsulated towards the center of the bead. (**d**) Bead with a hole

9. Slowly expel the FA/follicle from the tip (at an ~45° angle over the thrombin/Ca²⁺ solution), so it hangs like a drop from the end of the tip (Fig. 3b). Very gently touch the very bottom of the drop to the thrombin/Ca²⁺ solution to form FA-IPN bead (*see* **Note 19**). Watch the position of the encapsulated follicles (Fig. 4b, c) before the beads are fully cross-linked and turn opaque (*see* **Note 20**). Sometimes the beads may be empty (Fig. 4a) as follicles may get stuck in the pipette tips. If the pipette tip touches the cross-linking solution while dropping the bead, the bead may end up with a hole (Fig. 4d).

10. Allow the bead to cross-link for 2 min (*see* **Note 21**). In the meantime, repeat **steps 3–5** with the remaining follicles. Transfer fully cross-linked beads (Fig. 3c) to the 60 × 10 mm dish containing pre-equilibrated MM using a pair of curved forceps (Fig. 3d) (*see* **Note 22**).

11. Use a new 10 μL tip for each encapsulation—this prevents follicles from sticking to the used tip and the possibility of calcium blocking the tip that accidentally comes in touch with the

cross-linking solution. Also, changing the tips will help prevent potential cross-contamination during encapsulation and will help maintain a healthy culture.

12. Alternate between the IVF dishes with isolated follicles. Keep only 1 IVF dish outside the incubator for no longer than 10 min (*see* **Note 23**). Even though not all of the follicles from a specific IVF dish will be encapsulated in one round, put the dish back to the incubator to let the MM re-equilibrate. In the meantime, move on to a different dish to continue encapsulation. Repeat Subheading 3.2, **steps 5–11**, until all the isolated follicles have been encapsulated and all the beads are transferred to the 60 × 10 mm dish with pre-equilibrated MM (*see* **Note 24**). Count the number of beads and write the number on the lid.

3.3 Ovarian Follicle Culture

1. Mouse ovarian follicles are cultured in αMEM-based GM. To make GM, supplement αMEM (*see* **Note 24**) with 10 mg/mL fetuin, 3 mg/mL BSA, 5 μg/mL insulin, 5 μg/mL transferrin, and 5 ng/mL selenium. Then, sterile filter the solution using 0.2 μm cellulose acetate syringe filter. Add 10 mIU/mL rhFSH and invert gently to mix (*see* **Notes 25–27**).

2. To prepare aprotinin aliquots, calculate the amount of aprotinin using the information provided by the supplier (*see* **Note 28**). According to the total amount in the container, dissolve all the powder in sterile DPBS to obtain a concentration of 1 TIU/mL. Aliquot this stock solution as 50 or 100 μL aliquots and store at −20 °C (*see* **Notes 29** and **30**).

3. Count the number of beads by summing the numbers from Subheading 3.2, **step 12**. Based on the total number of the follicles, calculate the desired volume of GM: V (GM) = the number of follicles × 100 μL × 105% (5% more for evaporation). Equilibrate the GM in a dish. Prepare 96-well plates as in Fig. 5. The center wells of the plate should be filled with GM (100 μL/well), five wells in each row, three rows in total. All the other wells should be filled with sterile DPBS (100 μL/well) to minimize evaporation. Transfer the plates into the incubator to equilibrate.

4. After all the plates are equilibrated, use curved forceps to carefully transfer one bead to each well of GM in the 96-well plates. Every 2 days, after completing the imaging of the cultured follicles (*see* Subheading 3.5), replace half (50 μL) of the GM in the well with fresh, pre-equilibrated GM.

3.4 Optimal Aprotinin Concentration Determination

The degraded fibrin appears as an optically clear ring around the encapsulated follicle, which can be used as a surrogate for toxicant screening. To employ the optical clearance due to fibrin degradation, we have identified the range of aprotinin that maximized the

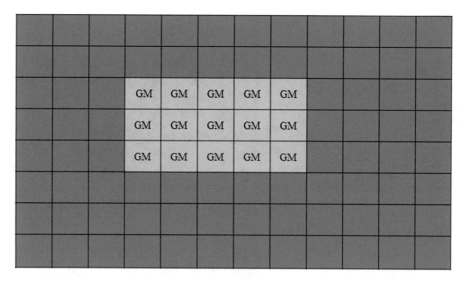

Fig. 5 96-Well plate setup with DPBS (blue) on the outer side as humidity control and GM (green) in the middle

control of fibrin degradation without affecting follicle growth. Suboptimal concentrations of aprotinin (less than 0.01 TIU/mL) resulted in uncontrolled fibrin degradation before ovotoxic effects could take place [19, 26]. Since fibrin gels contract in vitro [27], concentrations above 0.1 TIU/mL additionally lead to follicle death due to the inability to degrade fibrin and expand in condensed FA-IPN [19]. For the 12.5 mg/mL fibrin concentration in the FA-IPN, optimal aprotinin concentration is between 0.025 and 0.05 TIU/mL [26]. Aprotinin concentrations in this range slow down fibrin degradation and allow fibrin to be present at later time points to serve as the surrogate reporter for follicle health.

3.5 Follicle Imaging and Analyses

Image the cultured follicles every 2 days before changing the media. Turn on the imaging microscope and set up the imaging software. Position the first 96-well plate with follicles onto the microscope. First, use a lower magnification (e.g., 5×) to find the first bead in the first row. Then, record and measure the fibrin degradation ring around the follicle at 5×. Once you locate the bead in the well, switch to a higher magnification (e.g., 20×) to acquire an image of the follicle. Proceed to image all the follicles in the plate. Make sure that the imaging does not take longer than 10 min; otherwise, place the plate back to the incubator and return to it 30 min later to let the GM re-equilibrate (*see* **Note 31**).

Follicle growth in FA-IPN can be monitored via imaging every other day (Fig. 6). Follicle morphology can be examined with a 20× objective (Fig. 6a–d). To quantitatively measure follicle growth in FA-IPN, draw two orthogonal lines using ImageJ from basement membrane to basement membrane across the oocyte

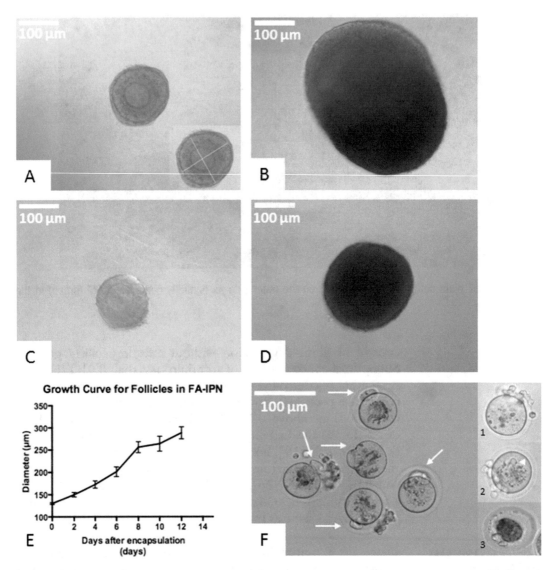

Fig. 6 Follicles in FA-IPN with aprotinin-controlled degradation are used to screen for toxicants in vitro. (**a** and **c**) Representative images of follicles in FA-IPN on D2. Insert of **a** shows the measurements of follicle diameter: 2 orthogonal lines across the oocyte, excluding the ECM. (**b** and **d**) Representative D10 images of live (**b**) and dead (**d**) follicles. (**e**) Follicle growth curve in FA-IPN. Data are presented as mean ± SEM, with n = 30. (**f**) Representative image of in vitro follicle maturation (IVFM). Arrows: MII oocytes. Arrowhead: Germinal vesicle

(Fig. 6a insert), calculate the average of the two measurements for each follicle, and generate a growth curve (Fig. 6e). Once the follicles reach 300 μm in diameter [28], in vitro follicle maturation (IVFM) can be performed to evaluate oocyte quality [17]. The oocytes were considered to be in metaphase I (Fig. 6f-1), if neither the germinal vesicle (Fig. 6f-2) nor the first polar body was visible. If a polar body was present in the perivitelline space, the oocytes were classified as metaphase II (Fig. 6f). Fragmented or shrunken

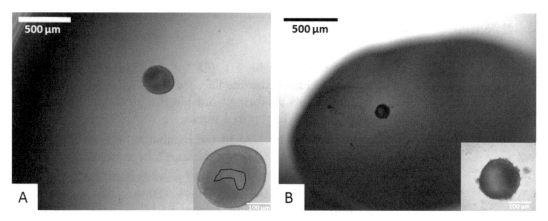

Fig. 7 Representative day 6 (D6) images of FA-IPN application for toxicology test (12.5 nM DMBA). (**a**) Untreated growing follicle in the control condition, antrum outlined. (**b**) Treated and dead follicle with minimal fibrin degradation

oocytes were classified as degenerated and were discarded (Fig. 6f-3). Healthy, competent oocytes from follicles cultured in FA-IPN should resume meiosis to metaphase II (MII) stage with one polar body present in the perivitelline space.

3.6 Follicles in FA-IPN for Toxicology Tests

We have demonstrated how the FA-IPN could be applied to test the ovotoxic effects of an anticancer drug doxorubicin (DXR) [26]. Follicles cultured in FA-IPN with aprotinin-controlled fibrin degradation can serve for in vitro screening of environmental toxicants such as 7,12-dimethylbenz[a]anthracene (DMBA) that has been previously shown to be ovotoxic [29]. To evaluate the effects of DMBA in our system, DMBA was dissolved in DMSO as a stock solution and was then further diluted with GM to reach a final concentration of 75 and 12.5 nM based on previous report [29]. The final concentration of DMSO in GM was kept at 0.5% (v/v) [26]. Follicles were treated with DMBA in GM on days 0, 2, and 4. When treating follicles encapsulated in FA-IPN with 12.5 nM DMBA, follicles affected by DMBA exposure showed little fibrin degradation (Fig. 7b) compared to untreated healthy follicles in the control condition (Fig. 7a). Quantification of fibrin degradation could also be applied as previously reported [26]. To confirm our findings using the fibrin degradation as the surrogate reporter, we also examined the follicle morphology at a higher magnification. Indeed, follicles reached smaller sizes, presented darker colors, and had no antrum formations (Fig. 7b insert). Therefore, fibrin degradation area can serve as a surrogate reporter for follicle health.

4   Notes

1. DM is buffered with HEPES for the atmospheric CO_2 level and is therefore the media used for all the procedures and manipulations performed outside the incubator. DM can be stored at 4 °C for up to 4 weeks.

2. Do not warm up DM in the incubator because the HEPES cannot buffer for CO_2 and the media becomes acidic.

3. MM is the media used to keep the isolated follicles in the incubator for all the procedures after isolation and before the transfer into GM. MM can be stored at 4 °C for up to 4 weeks.

4. Try to keep follicles separated from each other in the media at all times to avoid them sticking to each other. If indeed follicles aggregate, you can use the same approach with the needles to pull them apart for individual follicle encapsulation.

5. To minimize evaporation of FA droplets, follicle encapsulation should be done at room temperature. If you are performing follicle isolation on a heated stage, make sure that you turn off the heating prior to follicle encapsulation.

6. Sterile alginate aliquots should be prepared ahead of time. The aliquots should be stored in −20 °C until use.

7. Aqueous solution of alginate at concentrations greater than 0.2% (w/v) results in a viscous, difficult-to-filter solution.

8. It is important to prepare the alginate solution the day before encapsulation experiments, because the dissolution of alginate is slow, and this step can be time consuming.

9. If Tisseel sterile kit is unavailable, this is an alternative method to purify fibrinogen from other commercial sources such as Sigma.

    (a) Dissolve fibrinogen powder in MilliQ H_2O to reach an approximate concentration of 50 mg/mL. The supernatant contains the desired fibrinogen solution while the pellet is denatured protein or other particles, and thus transfer the supernatant into new tubes.

    (b) While fibrinogen is dissolving, prepare the appropriate amount of TBS for dialysis. Each dialysis tube should comfortably hold up to 25 mL of fibrinogen solution and needs to be dialyzed versus 4 L of TBS. For example, with a flat width of 40 mm, cut the tubes to be ~30 cm long. Once the number of dialysis bags is determined, make a 4 L solution of TBS for each dialysis bag by following the recipe below and titrate the pH to 7.4:

       -   17.44 g Tris–HCL
       -   2.56 g Tris base

- 32.00 g NaCl
- 0.80 g KCl

(c) Prepare dialysis tubing (M.W. = 6000–8000 Da), making sure to wet in TBS to separate. Wash each bag briefly with TBS in order to make sure that there are no holes. Each dialysis tube can comfortably hold up to 25 mL of fibrinogen solution and needs to be dialyzed in 4 L of TBS. Clip one end of the dialysis tubing on each bag and hold bag half inside large TBS beaker.

(d) Carefully transfer fibrinogen solution into bags and clip top of bags so that no solution is allowed to escape. Allow this closed dialysis bag to float in the TBS solution. Cover the beaker to prevent evaporation.

(e) Place all TBS beakers with dialysis bags on top of stirring plates (stir at slow-to-medium speeds) and allow the dialysis to occur overnight (at least 6 h).

(f) Carefully remove fibrinogen solution from dialysis bags and filter through 5.0 μm PVDF filter (EMD Millipore, Billerica, MA) to remove crude impurities.

(g) Measure the concentration of the fibrinogen solution after combining all the solutions using the NanoDrop 1000 spectrophotometer. Dilute the fibrinogen to desired concentration with sterile 25 mM TBS + 140 mM NaCl buffer if necessary.

(h) Sterile filter the diluted fibrinogen solution through 0.22 μm PES membrane filter (EMD Millipore, Billerica, MA).

(i) Aliquot fibrinogen solution into sterile 1.5 mL microcentrifuge tubes and store at −20 °C for short-term storage (up to 1 year) and −80 °C for long-term storage.

10. Sterile fibrinogen and thrombin aliquots should be prepared ahead of time. Both solutions are stored at −20 °C until use.

11. The osmolality of this solution should be 0.3624 Osm based on calculation.

12. Avoid leaving isolated follicles in MM for longer than 2 h: follicles tend to stick to the plastic. If follicles stuck to the plastic, try to lift the follicle from the bottom of the dish by pipetting up and down. Avoid excessive shear forces which may lead to basement membrane damage.

13. Concentrations of the encapsulating matrix can be adjusted to fit a specific application by changing the concentrations of alginate in the range between 0.125 and 0.8%, fibrinogen 1 and 75 mg/mL, and thrombin 1 and 500 mg/mL. For example, if a hydrogel of interest is with final concentrations of 0.5% algi-

nate and 12.5 mg/mL fibrinogen, mix 50 mg/mL fibrinogen with 1× DPBS and 1% alginate at 1:1:2 ratios.

14. Avoid vortexing this mixture: fibrinogen can get foamy. Pipet up and down in order to mix.

15. Practice making empty beads without follicles to optimize the angle and speed of extrusion.

16. This step is very critical to obtaining follicles for a successful study. Only healthy-appearing and undamaged follicles should be encapsulated. Keep multiple IVF dishes so you can alternate them in and out of the incubator to avoid lengthy exposure to ambient conditions.

17. Excessive remaining media can affect the cross-linking of FA-IPN, because of the dilution effect. Minimize the media uptake when transferring follicles.

18. Keep an eye on the wash droplet: change as needed if it gets pink due to the remaining media when transferring follicles.

19. When touching the calcium solution with the FA droplet, avoid the tip of the pipette touching the solution to prevent leaving a "hole" in the final FA-IPN bead (Fig. 4d).

20. To help better keep the follicles towards the center of the beads (Fig. 4), practice with some less than ideal follicles first to adjust how much of the 7.5 μL to take before taking up the target follicles.

21. The 50 mM concentration of $CaCl_2$ causes very fast gelation of the alginate. For a 7.5 μL bead, exposure to cross-linking solution for 2 min is sufficient for complete gelation.

22. When multiple FA-IPNs are present in the thrombin/Ca^{2+} cross-linking solution, highly cross-linked beads appear more opaque than those that are less cross-linked (Fig. 3c). This optical difference may help in identifying FA-IPNs for transferring to MM (Fig. 3d) to avoid extended exposure of follicles to Ca^{2+} ions.

23. To avoid drastic pH changes, limit the exposure of αMEM-based media such as MM and GM to ambient air for no longer than 10 min.

24. It is necessary to leave FA-IPNs in the MM for at least 30 min before plating the beads into GM: this process allows equilibration of the alginate matrix with the media.

25. The αMEM used for making GM should be opened in less than a month.

26. GM can be stored at 4 °C for up to 2 weeks.

27. If follicle isolation process is performed on a laboratory bench and/or the tissue used is obtained in an unsterile way, directly

from an abattoir for example, add 0.5% (v/v) Pen/Strep to GM to prevent potential contamination during the culture.

28. For example, for 3–7 TIU/(mg solid), apply an average of 5.5 TIU/(mg solid) to calculate the total amount of aprotinin.

29. Aprotinin aliquots can be stored at −20 °C for short-term or −80 °C for long-term storage.

30. Do not refreeze aprotinin solution. Use a fresh aliquot every time.

31. Plates with GM should not stay out of the incubator for more than 10 min at a time. Preferably, finish imaging all the follicles at one time to minimize the pH changes that are damaging to the cultured follicles. Adjust the number of wells with GM based on how long it takes to complete the imaging of one plate.

Acknowledgments

This work is supported by Reproductive Science Program (RSP) institutional funding (NIH U046944) from the University of Michigan to AS.

References

1. Xu Y, Duncan FE, Xu M, Woodruff TK (2015) Use of an organotypic mammalian in vitro follicle growth assay to facilitate female reproductive toxicity screening. Reprod Fertil Dev. doi:https://doi.org/10.1071/rd14375

2. Rotroff DM, Dix DJ, Houck KA, Knudsen TB, Martin MT, McLaurin KW, Reif DM, Crofton KM, Singh AV, Xia M, Huang R, Judson RS (2013) Using in vitro high throughput screening assays to identify potential endocrine-disrupting chemicals. Environ Health Perspect 121(1):7–14. https://doi.org/10.1289/ehp.1205065

3. Peretz J, Flaws JA (2013) Bisphenol A down-regulates rate-limiting Cyp11a1 to acutely inhibit steroidogenesis in cultured mouse antral follicles. Toxicol Appl Pharmacol 271(2):249–256. https://doi.org/10.1016/j.taap.2013.04.028

4. Cortvrindt RG, Smitz JEJ (2002) Follicle culture in reproductive toxicology: a tool for in-vitro testing of ovarian function? Hum Reprod Update 8(3):243–254. https://doi.org/10.1093/humupd/8.3.243

5. Lenie S, Smitz J (2009) Steroidogenesis-disrupting compounds can be effectively studied for major fertility-related endpoints using in vitro cultured mouse follicles. Toxicol Lett 185(3):143–152. https://doi.org/10.1016/j.toxlet.2008.12.015

6. Xu M, West-Farrell ER, Stouffer RL, Shea LD, Woodruff TK, Zelinski MB (2009) Encapsulated three-dimensional culture supports development of nonhuman primate secondary follicles. Biol Reprod 81(3):587–594. https://doi.org/10.1095/biolreprod.108.074732

7. O'Brien MJ, Pendola JK, Eppig JJ (2003) A revised protocol for in vitro development of mouse oocytes from primordial follicles dramatically improves their developmental competence. Biol Reprod 68(5):1682–1686. https://doi.org/10.1095/biolreprod.102.013029

8. Hirao Y, Nagai T, Kubo M, Miyano T, Miyake M, Kato S (1994) In vitro growth and maturation of pig oocytes. J Reprod Fertil 100(2):333–339. https://doi.org/10.1530/jrf.0.1000333

9. McLaughlin M, Telfer EE (2010) Oocyte development in bovine primordial follicles is promoted by activin and FSH within a two-step serum-free culture system. Reproduction 139(6):971–978. https://doi.org/10.1530/rep-10-0025

10. Spears N, Boland NI, Murray AA, Gosden RG (1994) Mouse oocytes derived from in vitro grown primary ovarian follicles are fertile. Hum Reprod 9(3):527–532

11. Xiao S, Zhang J, Romero MM, Smith KN, Shea LD, Woodruff TK (2015) In vitro follicle growth supports human oocyte meiotic maturation. Sci Rep 5:17323. https://doi.org/10.1038/srep17323

12. Silva GM, Rossetto R, Chaves RN, Duarte AB, Araujo VR, Feltrin C, Bernuci MP, Anselmo-Franci JA, Xu M, Woodruff TK, Campello CC, Figueiredo JR (2015) In vitro development of secondary follicles from pre-pubertal and adult goats cultured in two-dimensional or three-dimensional systems. Zygote 23(4):475–484. https://doi.org/10.1017/s0967199414000070

13. Laronda MM, Duncan FE, Hornick JE, Xu M, Pahnke JE, Whelan KA, Shea LD, Woodruff TK (2014) Alginate encapsulation supports the growth and differentiation of human primordial follicles within ovarian cortical tissue. J Assist Reprod Genet 31(8):1013–1028. https://doi.org/10.1007/s10815-014-0252-x

14. Xu J, Xu M, Bernuci MP, Fisher TE, Shea LD, Woodruff TK, Zelinski MB, Stouffer RL (2013) Primate follicular development and oocyte maturation in vitro. Adv Exp Med Biol 761:43–67. https://doi.org/10.1007/978-1-4614-8214-7_5

15. Shea LD, Woodruff TK, Shikanov A (2014) Bioengineering the ovarian follicle microenvironment. Annu Rev Biomed Eng 16:29–52. https://doi.org/10.1146/annurev-bioeng-071813-105131

16. Brito IR, Lima IM, Xu M, Shea LD, Woodruff TK, Figueiredo JR (2014) Three-dimensional systems for in vitro follicular culture: overview of alginate-based matrices. Reprod Fertil Dev 26(7):915–930. https://doi.org/10.1071/rd12401

17. Xu M, Kreeger PK, Shea LD, Woodruff TK (2006) Tissue-engineered follicles produce live, fertile offspring. Tissue Eng 12(10):2739–2746. https://doi.org/10.1089/ten.2006.12.2739

18. Shikanov A, Xu M, Woodruff TK, Shea LD (2009) Interpenetrating fibrin-alginate matrices for in vitro ovarian follicle development. Biomaterials 30(29):5476–5485. https://doi.org/10.1016/j.biomaterials.2009.06.054

19. Shikanov A, Xu M, Woodruff TK, Shea LD (2011) A method for ovarian follicle encapsulation and culture in a proteolytically degradable 3 dimensional system. J Vis Exp 49. doi:https://doi.org/10.3791/2695

20. Shikanov A, Smith RM, Xu M, Woodruff TK, Shea LD (2011) Hydrogel network design using multifunctional macromers to coordinate tissue maturation in ovarian follicle culture. Biomaterials 32(10):2524–2531. https://doi.org/10.1016/j.biomaterials.2010.12.027

21. Stefansdottir A, Fowler PA, Powles-Glover N, Anderson RA, Spears N (2014) Use of ovary culture techniques in reproductive toxicology. Reprod Toxicol 49:117–135. https://doi.org/10.1016/j.reprotox.2014.08.001

22. Beers WH (1975) Follicular plasminogen and plasminogen activator and the effect of plasmin on ovarian follicle wall. Cell 6(3):379–386. https://doi.org/10.1016/0092-8674(75)90187-7

23. El-Sadi F, Nader A, Becker C (2013) Ovulation and regulation of the menstrual cycle. In: Textbook of clinical embryology. Cambridge University Press, Cambridge, p 38

24. Xu M, Fazleabas AT, Shikanov A, Jackson E, Barrett SL, Hirshfeld-Cytron J, Kiesewetter SE, Shea LD, Woodruff TK (2011) In vitro oocyte maturation and preantral follicle culture from the luteal-phase baboon ovary produce mature oocytes. Biol Reprod 84(4):689–697. https://doi.org/10.1095/biolreprod.110.088674

25. Xu J, Lawson MS, Yeoman RR, Molskness TA, Ting AY, Stouffer RL, Zelinski MB (2013) Fibrin promotes development and function of macaque primary follicles during encapsulated three-dimensional culture. Hum Reprod 28(8):2187–2200. https://doi.org/10.1093/humrep/det093

26. Zhou H, Malik MA, Arab A, Hill MT, Shikanov A (2015) Hydrogel based 3-dimensional (3D) system for toxicity and high-throughput (HTP) analysis for cultured murine ovarian follicles. PLoS One 10(10):e0140205. https://doi.org/10.1371/journal.pone.0140205

27. Huang YC, Dennis RG, Larkin L, Baar K (2005) Rapid formation of functional muscle in vitro using fibrin gels. J Appl Physiol (1985) 98(2):706–713. https://doi.org/10.1152/japplphysiol.00273.2004

28. Xiao S, Duncan FE, Bai L, Nguyen CT, Shea LD, Woodruff TK (2015) Size-specific follicle selection improves mouse oocyte reproductive outcomes. Reproduction 150(3):183–192. https://doi.org/10.1530/REP-15-0175

29. Madden JA, Hoyer PB, Devine PJ, Keating AF (2014) Acute 7,12-dimethylbenz[a]anthracene exposure causes differential concentration-dependent follicle depletion and gene expression in neonatal rat ovaries. Toxicol Appl Pharmacol 276(3):179–187. https://doi.org/10.1016/j.taap.2014.02.011

Chapter 6

Layer-by-Layer Engineered Polymer Capsules for Therapeutic Delivery

Rona Chandrawati

Abstract

Polymer capsules fabricated via layer-by-layer (LbL) assembly have emerged as promising carriers for therapeutic delivery. The versatile assembly technique allows an extensive choice of materials to be incorporated as constituents of the multilayers, which therefore endow capsules with specific properties and functionalities. This chapter describes protocols for fabrication of LbL-engineered poly(methacrylic acid) (PMA) capsules for applications in gene delivery, including (1) synthesis of building blocks, (2) cargo encapsulation, (3) multilayer film formation, (4) surface modification, and (5) cross-linking of multilayer films and dissolution of particle templates. DNA is adsorbed onto positively charged silica particle templates, followed by formation of polymer films via hydrogen-bonded multilayers of thiol-functionalized PMA and poly(N-vinylpyrrolidone) (PVP). The outer polymer membranes can be surface modified with copolymers of PMA and poly(ethylene glycol) (PEG). Upon film stabilization and dissolution of particle templates, disulfide-cross-linked DNA-loaded PMA capsules are obtained, which serve as therapeutic carriers that can degrade and facilitate cargo release in intracellular reducing environment.

Key words Layer-by-layer, Polymer capsules, Poly(methacrylic acid), Poly(ethylene glycol), DNA, Drug delivery, Redox-responsive, Disulfide bond, Surface conjugation

1 Introduction

Layer-by-layer (LbL) assembly technique [1–5] is one of the most rapidly growing means of fabricating functional multilayer films and is simply based on the sequential adsorption of complementary species. A vast number of materials can be used including polymers [6], proteins [7], nucleic acids [8], and nanoparticles [9], and multilayers can be deposited on planar and colloidal supports. Upon removal of sacrificial colloidal templates, hollow capsules are obtained. This assembly technique provides facile control over morphology, composition, and film thickness on nanometer scale, which in turn defines the permeability of the capsule membrane. Through the judicious choice of building blocks, multilayer films can be readily engineered with tailored properties and stimuli

Kanika Chawla (ed.), *Biomaterials for Tissue Engineering: Methods and Protocols*, Methods in Molecular Biology, vol. 1758, https://doi.org/10.1007/978-1-4939-7741-3_6, © Springer Science+Business Media, LLC, part of Springer Nature 2018

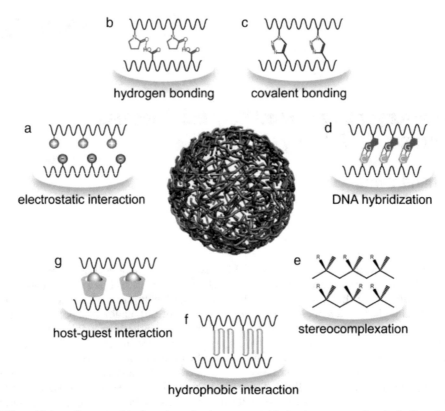

Fig. 1 Different interactions used to form layer-by-layer-assembled polymer capsules, including electrostatic interaction (**a**), hydrogen bonding (**b**), covalent bonding (**c**), DNA hybridization (**d**), stereocomplexation (**e**), hydrophobic interaction (**f**), and host-guest interaction (**g**). Reproduced from [6] with permission from Elsevier

responses. Owing to its simplicity and versatility, LbL assembly technique is of high interest in the biomaterials field.

Different interactions have been used to drive LbL assembly process (Fig. 1) [6]. Multilayers assembled from hydrogen-bonded systems allow the incorporation of uncharged polymers that otherwise cannot be used in electrostatic-based systems, and they can be engineered to exhibit desired stimuli-responsive properties at physiologically relevant conditions [10, 11]. Hydrogen-bonded polymer pairs based on poly(methacrylic acid) (PMA) and poly(N-vinylpyrrolidone) (PVP) serve as one of the prominent LbL-assembled capsule systems due to their high colloidal stability, (bio)degradability, and biocompatibility [12–16], and is therefore chosen as an example of LbL assembly system detailed in this chapter.

PMA (hydrogen bond donor) is sequentially layered to interact with PVP (hydrogen bond acceptor) on colloidal templates and the polymer pairs remain stable at pH values below the pKa of PMA (below pH 6.5) when PMA is protonated. Above the pKa of PMA (above pH 6.5), the polymer pairs will spontaneously disassemble due to deprotonation of carboxylic acid groups. Hence, to facilitate stabilization of the multilayers at physiological pH (pH 7.4), PMA is functionalized with thiol moieties (PMA$_{SH}$). Upon completion of PMA$_{SH}$/PVP deposition steps, the thiol groups on the polymer chains are converted into bridging disulfide linkages and the resulting cross-linked polymer capsules remain stable at physiological pH [16–18].

(Bio)degradable disulfide-cross-linked PMA capsules have been used to encapsulate a range of low-molecular-weight species and macromolecules, including DNA [19–21], lipophilic drugs [15], antigenic peptides [13, 22], intact proteins [14], and liposomes [23–26]. These capsules can be deconstructed in response to intracellular glutathione, which serve as a biological trigger to facilitate controlled and selective cargo release under intracellular reducing environment [16, 27]. The polymer shells can be modified to adapt to a range of applications; for example they can be functionalized with poly(ethylene glycol) (PEG), a bioinert polymer to endow surfaces with low-fouling properties, which can reduce interactions of drug carriers with phagocytes and prolong their circulation time in the bloodstream before they reach the target sites [28–30].

This chapter describes protocols for fabrication of (bio)degradable DNA-loaded PMA capsules for applications in intracellular gene delivery, including (1) synthesis of building blocks, (2) cargo encapsulation, (3) multilayer film formation, (4) surface modification, and (5) cross-linking of multilayer films and dissolution of particle templates (Fig. 2). Common centrifugation technique is employed for the LbL assembly method.

2 Materials

2.1 Chemicals and Reagents

Store chemicals and reagents at room temperature unless otherwise stated. For chemicals stored at 4 and −20 °C, always equilibrate to room temperature before opening. Be sure to read MSDS for each chemical before use.

1. Poly(methacrylic acid) (PMA, MW 15,000 Da, 30 wt% solution) (Polysciences).

2. Poly(N-vinylpyrrolidone) (PVP, MW 10,000 Da) (Sigma-Aldrich).

3. Pyridine dithioethylamine hydrochloride (PDA) (3A SpeedChemical): Store at −20 °C.

4. 1-Ethyl-3-(3-dimethylaminopropyl)carbodiimide hydrochloride (EDC) (Thermo Scientific): Store at −20 °C.

5. Dithiothreitol (DTT) (Sigma-Aldrich): Store at 4 °C.

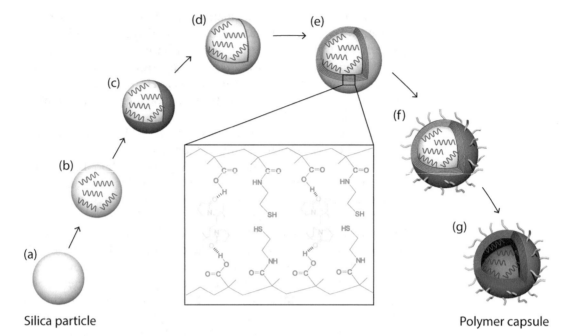

Silica particle   Polymer capsule

Fig. 2 Schematic illustration of polymer capsule assembly via layer-by-layer technique. Positively charged silica particle template (**a**) is coated with negatively charged DNA (**b**) via electrostatic interaction, followed by sequential deposition of thiol-functionalized poly(methacrylic acid) (PMA_{SH}) (**c**) and poly(N-vinylpyrrolidone) (PVP) (**d**) via hydrogen bonding. Once the desired number of polymer bilayers is achieved (**e**), the outer polymer membrane can be surface modified with copolymers of poly(methacrylic acid) and poly(ethylene glycol) (PMA_{SH}-PEG) (**f**). Oxidation of the PMA_{SH} thiol groups into bridging disulfide linkages and removal of the sacrificial particle template results in (bio)degradable disulfide-cross-linked poly(methacrylic acid) capsule encapsulating DNA (**g**)

6. Alexa Fluor 488 maleimide (Molecular Probes): Dissolve 1 mg of Alexa Fluor 488 maleimide in 1 mL of anhydrous dimethyl sulfoxide (DMSO) and store aliquots of 10 μL at −20 °C.

7. Maleimide poly(ethylene glycol) (MAL-PEG, MW 5,000 Da) (JenKem Technology USA): Store at −20 °C.

8. 2,2′-Dithiodipyridine (Sigma-Aldrich): Store at 4 °C.

9. Aminated SiO_2 particles 1 μm diameter (5 wt% solution) (microParticles GmbH): Store at 4 °C.

10. 5-Carboxytetramethylrhodamine   (TAMRA)-labeled   DNA containing 15 repeating thymine (T) residues followed by 15 repeating cytosine (C) residues ($polyT_{15}C_{15}$) are custom synthesized by TriLink BioTechnologies: Store at −20 °C.

11. Hydrofluoric acid (HF) (Sigma-Aldrich). *Caution! HF is very toxic.*

12. Ammonium fluoride (NH_4F) (Sigma-Aldrich). *Caution! NH_4F is very toxic.*

2.2 Buffer Solutions

Prepare all solutions using ultrapure water (resistivity of 18 MΩ. cm at 25 °C) and store at room temperature. Pass all buffer solutions through a 0.2 μm filter before use.

1. 10 mM Phosphate-buffered saline (PBS), pH 7.4.

2. 20 mM Sodium acetate (NaOAc) buffer, pH 4.0.

3. 20 mM 3-(N-morpholino)propanesulfonic acid (MOPS) buffer, pH 8.0.

4. 10 mM Tris–HCl, 1 mM EDTA (Tris–EDTA buffer), pH 7.5.

2.3 Equipment and Consumables

1. Analytical balance.

2. Vortex mixer.

3. Centrifuge.

4. P1000, P200, P20 pipettes and tips.

5. Freeze-dryer/lyophilizer.

6. Flow cytometer (e.g., CyFlow Space Partec GmbH).

7. Fluorescence microscope (e.g., Olympus IX73 inverted microscope with a 60× oil immersion objective and fluorescence filter sets).

8. Dialysis tubing cellulose membrane, MWCO 12,400 Da, flat width 32 mm (Sigma-Aldrich).

9. Dialysis tubing closures, size 70 mm (Sigma-Aldrich).

10. Microsep™ centrifugal devices, MWCO 10,000 Da (Pall Corporation).

11. 1.5 mL tubes (Eppendorf).

3 Methods

These protocols can be performed at room temperature on a laboratory bench unless otherwise stated. Take appropriate laboratory safety measures and always use personal protective equipment (safety glasses, gloves, and lab coat) in a laboratory environment.

3.1 Synthesis of Thiol-Functionalized Poly(methacrylic acid) (PMA_{SH})

1. Weigh 300 mg of PMA into a glass vial.

2. Dissolve 64.6 mg of EDC in 3.2 mL of PBS and 44.5 mg of PDA in 2.2 mL of PBS in two separate glass vials (see **Note 1**).

3. While vortexing, add EDC solution to the PMA solution and continue mixing for 15 min.

4. After 15 min, add PMA/EDC mixture to the PDA solution (see **Note 2**).

5. Incubate the reaction overnight with continuous mixing.

6. Carefully transfer the reaction mixture into a dialysis tubing cellulose membrane MWCO 12,400 Da (*see* **Note 3**). Ensure that the dialysis tube is tightly sealed at both ends with dialysis tubing closures. Purify the polymer solution by extensive dialysis against 4 L of ultrapure water in a beaker for 2 days. Exchange water daily during this period.

7. Carefully transfer the purified polymer solution from the dialysis tube into a glass vial. Freeze the vial containing the polymer solution at −80 °C for 3–4 h.

8. Freeze-dry the polymer solution to obtain a white powder of PDA-functionalized PMA (PMA$_{PD}$). Store at 4 °C until further use.

9. To obtain PMA$_{SH}$, incubate 1 mg of PMA$_{PD}$ with 10 μL of 0.5 M DTT in MOPS buffer at 37 °C for at least 15 min (*see* **Note 4**). Dilute the polymer solution with 490 μL of NaOAc buffer to obtain 2 mg/mL PMA$_{SH}$ solution (*see* **Note 5**).

3.2   Preparation of Alexa Fluor 488-Labeled Thiol-Functionalized Poly(methacrylic acid) (PMA$_{SH}$-488)

1. Incubate 10 mg of PMA$_{PD}$ with 100 μL of 0.5 M DTT in MOPS buffer at 37 °C for at least 15 min (*see* **Note 4**). Dilute the polymer solution with 900 μL of PBS to obtain 10 mg/mL PMA$_{SH}$ solution.

2. Carefully transfer the reaction mixture into a Microsep™ centrifugal device MWCO 10,000 Da and remove excess DTT by three centrifugation/reconstitution cycles (7500 × *g* for 10 min each cycle) with PBS (*see* **Note 6**). Remove the supernatant and reconstitute the concentrate with 1 mL of PBS in each cycle.

3. Add 10 μL of 1 mg/mL Alexa Fluor 488 maleimide in DMSO to the PMA$_{SH}$ solution. Mix well and incubate the reaction for 2 h with continuous mixing (*see* **Note 7**).

4. Remove excess Alexa Fluor 488 maleimide by five centrifugation/reconstitution cycles (7500 × *g* for 10 min each cycle) with ultrapure water. Remove the supernatant and reconstitute the concentrate with 1 mL of ultrapure water in each cycle.

5. Carefully transfer the polymer solution into a glass vial. Freeze the vial containing the polymer solution at −80 °C for 1 h.

6. Freeze-dry the polymer solution to obtain the powder of PMA$_{PD}$-488. Store at 4 °C until further use (*see* **Note 7**).

7. To obtain PMA$_{SH}$-488, incubate 1 mg of PMA$_{PD}$-488 with 10 μL of 0.5 M DTT in MOPS buffer at 37 °C for at least 15 min. Dilute the polymer solution with 490 μL of NaOAc buffer to obtain 2 mg/mL PMA$_{SH}$-488 solution.

3.3 Preparation of Thiol-Functionalized Poly(methacrylic acid)-Poly(ethylene glycol) Conjugates (PMA$_{SH}$-PEG)

1. Incubate 10 mg of PMA$_{PD}$ with 100 µL of 0.5 M DTT in MOPS buffer at 37 °C for at least 15 min (*see* **Note 4**). Dilute the polymer solution with 900 µL of PBS to obtain 10 mg/mL PMA$_{SH}$ solution.

2. Carefully transfer the reaction mixture into a Microsep™ centrifugal device MWCO 10,000 Da and remove excess DTT by three centrifugation/reconstitution cycles (7500 × g for 10 min each cycle) with PBS (*see* **Note 6**). Remove the supernatant and reconstitute the concentrate with 1 mL of PBS in each cycle.

3. Add 800 µL of 50 mg/mL MAL-PEG in PBS to the PMA$_{SH}$ solution. Mix well and incubate the reaction for 2 h with continuous mixing.

4. Remove unreacted MAL-PEG by five centrifugation/reconstitution cycles (7500 × g for 10 min each cycle) with ultrapure water. Remove the supernatant and reconstitute the concentrate with 1 mL of ultrapure water in each cycle.

5. Carefully transfer the polymer solution into a glass vial. Freeze the vial containing the polymer solution at −80 °C for 1 h.

6. Freeze-dry the polymer solution to obtain powder of PMA$_{PD}$-PEG. Store at 4 °C until further use.

7. To obtain PMA$_{SH}$-PEG, incubate 1 mg of PMA$_{PD}$-PEG with 10 µL of 0.5 M DTT in MOPS buffer at 37 °C for at least 15 min. Dilute the polymer solution with 490 µL of NaOAc buffer to obtain 2 mg/mL PMA$_{SH}$-PEG solution (*see* **Note 8**).

3.4 Loading of Therapeutic Cargo onto Particle Templates

1. Prepare 250 µL of 1 µm diameter aminated SiO$_2$ particles (5 wt% solution) in an Eppendorf tube.

2. Wash the SiO$_2$ particles by three centrifugation/redispersion cycles (1000 × g for 30 s each cycle) with Tris–EDTA buffer. Remove the supernatant and redisperse the particles with 250 µL of Tris–EDTA buffer in each cycle. Vortex the particle solution for 30 s between centrifugation cycles and finally disperse the particles with 125 µL of Tris–EDTA buffer in the last step (*see* **Note 9**).

3. Add 125 µL of 1 µM TAMRA-polyT$_{15}$C$_{15}$ in Tris–EDTA buffer into the particle solution. Vortex the particle mixture for 30 s and allow the adsorption to proceed for 30 min with constant shaking (*see* **Notes 10** and **11**).

4. Wash the coated particles by two centrifugation/redispersion cycles (1000 × g for 30 s each cycle) with Tris–EDTA buffer and further two centrifugation/redispersion cycles (1000 × g for 30 s each cycle) with NaOAc buffer (*see* **Note 12**). Remove the supernatant and redisperse the particles with 250 µL of buffer in each cycle. Vortex the particle solution for 30 s

between centrifugation cycles and finally disperse the coated particles with 125 μL of NaOAc buffer in the last step for the subsequent assembly of polymer layers (*see* **Note 9**).

3.5 Layer-by-Layer Deposition of Polymers

1. Add 125 μL of 2 mg/mL PMA_{SH} or PMA_{SH}-488 in NaOAc buffer into the DNA-coated particle solution (*see* **Note 13**). Vortex the particle mixture for 30 s and allow the adsorption to proceed for 15 min with constant shaking (*see* **Note 11**).

2. Wash the polymer-coated particles by three centrifugation/ redispersion cycles ($1000 \times g$ for 30 s each cycle) with NaOAc buffer. Remove the supernatant and redisperse the particles with 250 μL of NaOAc buffer in each cycle. Vortex the particle solution for 30 s between centrifugation cycles and disperse the coated particles with 125 μL of NaOAc buffer in the last step.

3. For the adsorption of subsequent polymer layer, add 125 μL of 2 mg/mL PVP in NaOAc buffer into the particle solution. Vortex the particle mixture for 30 s and allow the adsorption to proceed for 15 min with constant shaking (*see* **Note 11**).

4. Wash the polymer-coated particles by three centrifugation/ redispersion cycles ($1000 \times g$ for 30 s each cycle) with NaOAc buffer. Remove the supernatant and redisperse the particles with 250 μL of NaOAc buffer in each cycle. Vortex the particle solution for 30 s between centrifugation cycles and disperse the coated particles with 125 μL of NaOAc buffer in the last step.

5. Repeat **steps 1–4** Sec. 3.5 until five bilayers of PMA_{SH}/PVP or PMA_{SH}-488/PVP are assembled onto the particles (*see* **Note 14**). The multilayer buildup can be analyzed using a flow cytometer according to the manufacturer's instructions for the instrument used, and confirmed by a progressive increase in the fluorescence of the particles.

If no additional surface modification is required, proceed to Subheading 3.7. Otherwise, proceed to the next step.

3.6 Surface Modification

1. To modify the outer surface of the polymer multilayers with PEG, add 125 μL of 2 mg/mL PMA_{SH}-PEG in NaOAc buffer into the particle solution (*see* **Note 15**). Vortex the particle mixture for 30 s and allow the adsorption to proceed for 15 min with constant shaking.

2. Wash the PEG-coated particles by three centrifugation/redispersion cycles ($1000 \times g$ for 30 s each cycle) with NaOAc buffer. Remove the supernatant and redisperse the particles with 250 μL of NaOAc buffer in each cycle. Vortex the particle solution for 30 s between centrifugation cycles and disperse the particles with 250 μL of NaOAc buffer in the last step.

3.7 Polymer Film Stabilization and Formation of Polymer Capsules

1. To cross-link the thiols within the polymer layers, add 250 μL of 1 mg/mL 2,2′-dithiodipyridine in NaOAc buffer (*see* **Note 16**). Vortex the particle mixture for 30 s and allow the adsorption to proceed overnight with constant shaking.

2. Wash the particles by three centrifugation/redispersion cycles (1000 × *g* for 30 s each cycle) with NaOAc buffer to remove unreacted 2,2′-dithiodipyridine. Remove the supernatant and redisperse the particles with 250 μL of NaOAc buffer in each cycle. Vortex the particle solution for 30 s between centrifugation cycles and finally disperse the particles with 250 μL of NaOAc buffer in the last step.

3. To obtain hollow polymer capsules, dissolve the SiO_2 cores by adding 250 μL of 2 M HF/8 M NH_4F (buffered HF) into the particle solution for 2 min with gentle agitation. *Caution! HF and NH_4F are highly toxic. Perform this step in a designated laboratory fume hood. Extreme care should be taken when handling HF and NH_4F solutions and only small quantities should be prepared* (*see* **Note 17**). Wash the particles by three centrifugation/redispersion cycles (4500 × *g* for 3 min each cycle) with NaOAc buffer (*see* **Note 18**). Carefully remove the supernatant from each centrifugation cycles, neutralize immediately with 5 M $CaCl_2$ solution, and redisperse the particles with 250 μL of NaOAc buffer in each cycle. Vortex the particle solution for 30 s between centrifugation cycles and finally disperse the particles with 250 μL of NaOAc buffer (*see* **Note 19**). The hollow polymer capsules can be imaged using a fluorescence microscope according to the manufacturer's instructions for the instrument used (Fig. 3).

4. For biological studies, the buffer solution used to disperse the hollow capsules can be exchanged to PBS simply by three centrifugation/redispersion cycles (4500 × *g* for 3 min each cycle) with the corresponding buffer (*see* **Note 20**). Flow cytometry can be used for absolute polymer capsule counting [31].

4   Notes

1. Freshly prepare EDC and PDA solutions before performing the reaction.

2. EDC reacts with carboxylic acid groups of PMA and an active intermediate, O-acylisourea, is formed. This intermediate is unstable in aqueous solutions and must react immediately with PDA.

3. Cut desired length of dialysis tubing and soak the whole tubing in ultrapure water for a few minutes to soften it for handling and to open the hollow tube. Keep the tubing moist throughout the experiment.

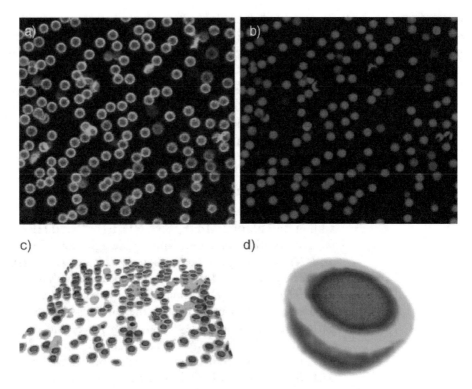

Fig. 3 Fluorescence microscopy images of disulfide-cross-linked poly(methacrylic acid) capsules encapsulating DNA (polyT$_{15}$C$_{15}$), showing fluorescence originating from the polymers labeled with Alexa Fluor 488 (green) (**a**) and DNA labeled with TAMRA (red) (**b**). Polymer capsules are monodisperse and DNA is uniformly distributed throughout the capsule interior. 3D cross-section reconstruction images of the polymer capsules are shown (**c, d**). Images (**a–c**) are 30 × 30 μm². The capsule in (**d**) is 1.5 μm in diameter. Reproduced from [12] with permission from Wiley-VCH

4. PMA$_{SH}$ should be fully dissolved in DTT prior to dilution with buffer solution.

5. This protocol yields PMA with 14 mol% of thiol modification. The degree of thiol functionalization of PMA can be characterized by measuring the absorbance of released PDA groups (λ_{max} = 343 nm) due to the addition of reducing agent DTT and quantified by correlation with a calibration curve of PDA.

6. DTT contains thiols and has to be removed thoroughly before initiating the maleimide reaction.

7. Cover the tube with aluminum foil to avoid exposure to light.

8. This protocol yields 6 mol% functionalization degree of PEG per PMA chain, with the remaining 8 mol% of thiol functionalization available for disulfide cross-linking.

9. Ensure that the SiO$_2$ particles are well dispersed before subsequent adsorption step.

10. Adsorption and loading of DNA are achieved through the electrostatic interaction between negatively charged polyT$_{15}$C$_{15}$ and positively charged SiO$_2$ particles.

11. To avoid aggregation, agitate particle solutions at all times during adsorption step.

12. Particles need to be dispersed in NaOAc buffer pH 4.0 to allow subsequent hydrogen-bonded multilayers of PMA_{SH} and PVP. At this pH, PMA_{SH} is protonated and acts as hydrogen bond donor.

13. Freshly prepare PMA_{SH} or PMA_{SH}-488 solution before performing LbL deposition.

14. No sonication is used at any step of assembly.

15. Freshly prepare PMA_{SH}-PEG solution before performing surface modification step.

16. Freshly prepare 2,2′-dithiodipyridine solution before performing cross-linking step.

17. Buffered HF solution (pH 5.0) is prepared by mixing 100 μL of 5 M HF and 150 μL of 13 M NH_4F. Preparation of HF/NH_4F solution and dissolution of SiO_2 particles must be performed in a designated fume hood. Neutralize all pipette tips immediately with 5 M $CaCl_2$ solution. The particle solution becomes clear upon addition of buffered HF solution.

18. Hollow polymer capsules require higher centrifugation speed and longer time to pellet ($4500 \times g$ for 3 min) than the core-shell particles ($1000 \times g$ for 30 s).

19. Hollow polymer capsules can be stored at 4 °C and are stable for up to 2 weeks.

20. Buffer solution to disperse the polymer capsules should be exchanged fresh just before use for subsequent studies.

Acknowledgments

RC acknowledges the support from the Australian Research Council Discovery Early Career Researcher Award (ARC DECRA DE170100068).

References

1. Richardson JJ, Björnmalm M, Caruso F (2015) Technology-driven layer-by-layer assembly of nanofilms. Science 348:aaa2491

2. Borges J, Mano JF (2014) Molecular interactions driving the layer-by-layer assembly of multilayers. Chem Rev 114:8883–8942

3. Hammond PT (2012) Building biomedical materials layer-by-layer. Mater Today 15:196–206

4. Städler B, Price AD, Zelikin AN (2011) A critical look at multilayered polymer capsules in biomedicine: drug carriers, artificial organelles, and cell mimics. Adv Funct Mater 21:14–28

5. Delcea M, Möhwald H, Skirtach AG (2011) Stimuli-responsive LbL capsules and nanoshells for drug delivery. Adv Drug Deliv Rev 63:730–747

6. Cui J, van Koeverden MP, Müllner M, Kempe K, Caruso F (2014) Emerging methods for the fabrication of polymer capsules. Adv Colloid Interf Sci 207:14–31

7. Lvov Y, Ariga K, Ichinose I, Kunitake T (1995) Assembly of multicomponent protein films by means of electrostatic layer-by-layer adsorption. J Am Chem Soc 117:6117–6123

8. Kato N, Lee L, Chandrawati R, Johnston APR, Caruso F (2009) Optically characterized DNA multilayered assemblies and phenomenological modeling of layer-by-layer hybridization. J Phys Chem C 113:21185–21195

9. Caruso F, Caruso RA, Möhwald H (1998) Nanoengineering of inorganic and hybrid hollow spheres by colloidal templating. Science 282:1111–1114

10. Such GK, Johnston APR, Caruso F (2011) Engineered hydrogen-bonded polymer multilayers: from assembly to biomedical applications. Chem Soc Rev 40:19–29

11. Kharlampieva E, Kozlovskaya V, Sukhishvili SA (2009) Layer-by-layer hydrogen-bonded polymer films: from fundamentals to applications. Adv Mater 21:3053–3065

12. Zelikin AN, Li Q, Caruso F (2006) Degradable polyelectrolyte capsules filled with oligonucleotide sequences. Angew Chem Int Ed 118:7907–7909

13. De Rose R, Zelikin AN, Johnston APR, Sexton A, Chong S-F, Cortez C, Mulholland W, Caruso F, Kent SJ (2008) Binding, internalization, and antigen presentation of vaccine-loaded nanoengineered capsules in blood. Adv Mater 20:4698–4703

14. Sexton A, Whitney PG, Chong S-F, Zelikin AN, Johnston APR, De Rose R, Brooks AG, Caruso F, Kent SJ (2009) A protective vaccine delivery system for *in vivo* T cell stimulation using nanoengineered polymer hydrogel capsules. ACS Nano 3:3391–3400

15. Sivakumar S, Bansal V, Cortez C, Chong S-F, Zelikin AN, Caruso F (2009) Degradable, surfactant-free, monodisperse polymer-encapsulated emulsions as anticancer drug carriers. Adv Mater 21:1820–1824

16. Zelikin AN, Li Q, Caruso F (2008) Disulfide-stabilized poly(methacrylic acid) capsules: formation, cross-linking, and degradation behavior. Chem Mater 20:2655–2661

17. Zelikin AN, Quinn JF, Caruso F (2006) Disulfide cross-linked polymer capsules: en route to biodeconstructible systems. Biomacromolecules 7:27–30

18. Chong S-F, Chandrawati R, Städler B, Park J, Cho J, Wang Y, Jia Z, Bulmus V, Davis TP, Zelikin AN, Caruso F (2009) Stabilization of polymer-hydrogel capsules via thiol-disulfide exchange. Small 5:2601–2610

19. Zelikin AN, Becker AL, Johnston APR, Wark KL, Turatti F, Caruso F (2007) A general approach for DNA encapsulation in degradable polymer microcapsules. ACS Nano 1:63–69

20. Price AD, Zelikin AN, Wang Y, Caruso F (2009) Triggered enzymatic degradation of DNA within selectively permeably polymer capsule microreactors. Angew Chem Int Ed 48:329–332

21. Price AD, Zelikin AN, Wark KL, Caruso F (2010) A biomolecular "ship-in-a-bottle": continuous RNA synthesis within hollow polymer hydrogel assemblies. Adv Mater 22:720–723

22. Chong S-F, Sexton A, De Rose R, Kent SJ, Zelikin AN, Caruso F (2009) A paradigm for peptide vaccine delivery using viral epitopes encapsulated in degradable polymer hydrogel capsules. Biomaterials 30:5178–5186

23. Chandrawati R, Städler B, Postma A, Connal LA, Chong S-F, Zelikin AN, Caruso F (2009) Cholesterol-mediated anchoring of enzyme-loaded liposomes within disulfide-stabilized polymer carrier capsules. Biomaterials 30:5988–5998

24. Chandrawati R, Hosta-Rigau L, Vanderstraaten D, Lokuliyana SA, Städler B, Albericio F, Caruso F (2010) Engineering advanced capsosomes: maximizing the number of subcompartments, cargo retention, and temperature-triggered reaction. ACS Nano 4:1351–1361

25. Chandrawati R, Odermatt PD, Chong S-F, Price AD, Städler B, Caruso F (2011) Triggered cargo release by encapsulated enzymatic catalysis in capsosomes. Nano Lett 11:4958–4963

26. Maina JW, Richardson JJ, Chandrawati R, Kempe K, van Koeverden MP, Caruso F (2015) Capsosomes as long-term delivery vehicles for protein therapeutics. Langmuir 31:7776–7781

27. Yan Y, Wang Y, Heath JK, Nice EC, Caruso F (2011) Cellular association and cargo release of redox-responsive polymer capsules mediated by exofacial thiols. Adv Mater 23:3916–3921

28. Chandrawati R, Chong S-F, Zelikin AN, Hosta-Rigau L, Städler B, Caruso F (2011) Degradation of liposomal subcompartments in PEGylated capsosomes. Soft Matter 7:9638–9646

29. Knop K, Hoogenboom R, Fischer D, Schubert US (2010) Poly(ethylene glycol) in drug delivery: pros and cons as well as potential alternatives. Angew Chem Int Ed 49:6288–6308

30. Cui J, De Rose R, Alt K, Alcantara S, Paterson BM, Liang K, Hu M, Richardson JJ, Yan Y, Jeffery CM, Price RI, Peter K, Hagemeyer CE, Donnelly PS, Kent SJ, Caruso F (2015) Engineering poly(ethylene glycol) particles for improved biodistribution. ACS Nano 9:1571–1580

31. Chandrawati R, Chang JYH, Reina-Torres E, Jumeaux C, Sherwood JM, Stamer WD, Zelikin AN, Overby DR, Stevens MM (2017) Localized and controlled delivery of nitric oxide to the conventional outflow pathway via enzyme biocatalysis: toward therapy for glaucoma. Adv Mater 29:1604932

Chapter 7

Controlling Fibrin Network Morphology, Polymerization, and Degradation Dynamics in Fibrin Gels for Promoting Tissue Repair

Erin P. Sproul, Riley T. Hannan, and Ashley C. Brown

Abstract

Fibrin is an integral part of the clotting cascade and is formed by polymerization of the soluble plasma protein fibrinogen. Following stimulation of the coagulation cascade, thrombin activates fibrinogen, which binds to adjacent fibrin(ogen) molecules resulting in the formation of an insoluble fibrin matrix. This fibrin network is the primary protein component in clots and subsequently provides a scaffold for infiltrating cells during tissue repair. Due to its role in hemostasis and tissue repair, fibrin has been used extensively as a tissue sealant. Clinically used fibrin tissue sealants require supraphysiological concentrations of fibrinogen and thrombin to achieve fast polymerization kinetics, which results in extremely dense fibrin networks that are inhibitory to cell infiltration. Therefore, there is much interest in developing fibrin-modifying strategies to achieve rapid polymerization dynamics while maintaining a network structure that promotes cell infiltration. The properties of fibrin-based materials can be finely controlled through techniques that modulate fibrin polymerization dynamics or through the inclusion of fibrin-modifying biomaterials. Here, we describe methods for characterizing fibrin network morphology, polymerization, and degradation (fibrinolysis) dynamics in fibrin constructs for achieving fast polymerization dynamics while promoting cell infiltration.

Key words Fibrin, Fibrinogen, Fibrin-based biomaterials, Tissue engineering, Biomaterials, Wound healing, Hemostasis, Clot formation

1 Introduction

Fibrin serves as the major structural protein in blood clots and plays a key role in promoting healing by providing physical support for infiltrating cells following vascular injury [1–3]. Fibrin is an endogenous, and therefore biocompatible, glycoprotein, which is degraded by existing fibrinolysis pathways [4]. Fibrin also contains numerous binding sites for cell surface receptors, growth factors, and extracellular matrix proteins [5, 6]. Due to fibrin's key role in hemostasis and wound healing, as well as its high level of

Kanika Chawla (ed.), *Biomaterials for Tissue Engineering: Methods and Protocols*, Methods in Molecular Biology, vol. 1758, https://doi.org/10.1007/978-1-4939-7741-3_7, © Springer Science+Business Media, LLC, part of Springer Nature 2018

bioactivity, fibrin-based constructs have been utilized extensively for wound repair applications. For example, fibrin has been utilized clinically as a wound sealant in fibrin glues and wound dressings [7–10]. Additionally, fibrin sealants are constructed in the form of a hydrogel, which relies on mild fabrication conditions, thereby allowing for cellularization via cell entrapment during gelation. Therefore, along with their utility in wound closure, fibrin sealants can be combined with cell therapies to further promote tissue repair. This can be achieved by using fibrin to deliver cells to damaged tissue [13] or by modifying fibrin constructs to direct specific cellular responses involved in wound repair. Fibrin constructs have been used to direct stem cell differentiation [11, 12], induce angiogenesis [14, 15], and stimulate extracellular matrix (ECM) production [16].

Clinically used fibrin tissue sealants require supraphysiological concentrations of fibrinogen and thrombin to achieve fast polymerization kinetics, which results in extremely dense fibrin networks that are inhibitory to cell infiltration. To create fibrin materials that promote wound repair while maintaining clinical utility, fibrin polymerization dynamics and structural features must be balanced in order to obtain a rapidly forming network with optimal network porosity. Therefore, there is much interest in developing fibrin-modifying materials to achieve rapid polymerization dynamics while maintaining a network structure that promotes cell infiltration. As an example, Suggs et al. demonstrated gels constructed from 10 mg/mL fibrinogen and 100 U/mL thrombin, which are higher than physiological levels but lower than levels used in most surgical sealants could be modified with PEG to obtain a gel that displayed fast polymerization dynamics while increasing MSC viability and promoting angiogenesis [13].

The phenomenon of fibrin polymerization is well characterized and the molecular details of this process can be exploited to obtain fibrin gels with optimized polymerization/degradation kinetics and network structure for promotion of wound repair [4]. Material properties of fibrin gels are inherently linked to polymerization; modification of fibrin polymerization dynamics directly affects the porosity, fiber thickness, and degree of branching of the polymerized gel, which in turn influences cellular infiltration [17, 18]. These network proprieties are also intrinsically linked to mechanical properties of the gel. Further modification of the mechanical properties of fibrin gels can be achieved via covalent cross-linking using ribose, lysyl oxidase, or factor XIII [19–21]. Alteration of fibrin polymerization at the molecular level has been achieved through a myriad of methods including alterations of pH, salt, and thrombin concentrations [22, 23]; incorporation of polyethylene glycol (PEG) and other polymers [24–26]; modification with colloids [27]; and incorporation of materials that directly interface with fibrin polymerization [28–33]. Given the diversity of available molecular techniques, it is not surprising that fibrin gelation results in the creation of biomaterials with a wide range of material properties. In this chapter, we describe methods for characterizing properties of fibrin networks for achieving fast polymerization dynamics while maintaining network structure sufficient for promoting cell infiltration; we focus on the analysis of fibrin network morphology via confocal microscopy and characterization of fibrin polymerization and degradation kinetics through absorbance and microscopy-based assays.

2   Materials

Prepare all solutions using ultrapure water (prepared by purifying deionized water to attain a sensitivity of 18 MΩ cm at 25 °C) and analytical grade reagents. Prepare and store all reagents at room temperature (unless indicated otherwise). A stock HEPES buffer solution at increased concentration (e.g., 10×) is later added, in order to maintain consistent salt and $CaCl_2$ concentrations between samples. This is important since variations in these components can affect final fibrin network properties. Once diluted, the final buffer concentration across samples is 25 mM HEPES, 150 mM NaCl, and 5 mM $CaCl_2$ pH 7.4. These concentrations can be modified for experimental needs, but maintaining consistency across experimental groups is important.

2.1   Fibrin Gel Components

1. Buffer: 25 mM HEPES, 150 mM NaCl, 5 mM $CaCl_2$ pH 7.4. (This can be prepared in a 10× stock concentration and added to achieve the final concentrations indicated.)

2. Fibrinogen: Human FIB3, plasminogen, von Willebrand factor, and fibronectin depleted (Enzyme Research Laboratories, South Bend, IN). Store at −80°C prepared in water (*see* **Notes 1** and **2**).

3. Thrombin: Human α-thrombin (Enzyme Research Laboratories, South Bend, IN) prepared in water (*see* **Notes 2** and **3**).

2.2   Confocal Microscopy

1. Confocal microscope with 63× oil immersion objective, Numerical Aperture (NA) = 1.4: e.g., Zeiss 510 VIS (*see* **Note 4**).

2. Fluorescently labeled fibrinogen: This can be purchased, e.g., Alexa Fluor 488 (Sigma, St. Louis, MO) or prepared through the use of a labeling kit according to the manufacturer's instructions (Thermo Fisher Scientific, Waltham, MA).

3. Glass microscope slides: 75 mm × 25 mm.

4. Glass coverslips: Thickness No. 1.5.

5. Double-sided tape: Or equivalent spacer to secure coverslip over fibrin gel onto slide.

6. Nail polish: To seal the coverslip edges.

2.3 Absorbance-Based Polymerization and Fibrinolysis Assays

1. Clear bottom 96-well plates (Nunc).

2. Multi-well pipets (5–50 μL and 50–300 μL).

3. Fibrin-modifying agent (experimental variable, e.g., fibrin-binding nanoparticle).

4. Plate reader, e.g., Synergy H4 Hybrid Multi-Mode Microplate Reader (BioTek, Winooski, VT).

5. Tissue plasminogen activator (tPA) (Millipore, Billerica, MA).

6. Plasminogen (Enzyme Research Laboratories, South Bend, IN).

7. Quant-iT protein assay (Thermo Fisher Scientific, Waltham, MA).

8. Plasmin (Enzyme Research Laboratories, South Bend, IN).

9. FXIIIa (Enzyme Research Laboratories, South Bend, IN).

2.4 Microscopy-Based Fibrinolysis Assays

1. Purified human fibrinogen: Human α-thrombin, e.g., FIB 3 (Enzyme Research Laboratories).

2. Fluorescently labeled fibrinogen, e.g., Alexa Flour-488 fibrinogen (Thermo Fisher Scientific, Waltham, MA).

3. Polydimethylsiloxane (PDMS) Elastomer Base and Crosslinker (Sylgard 184, Dow Corning).

4. Mold for casting with appropriate "Y" or "T" channel structure (Fig. 3).

5. Glass microscope slides.

6. Plasma cleaner, e.g., Basic Plasma Cleaner PDC-32G (Harrick Plasma, Ithaca, NY).

7. Small-diameter (<2 mm) biopsy punch or blunt-tipped needle.

8. Tape, high-vacuum grease, or tubing to seal device (Fig. 3).

3 Methods

Carry out all procedures at room temperature unless otherwise specified.

Overview of fibrin gel production for characterization of polymerization, structural, and degradation properties.

Fibrin hydrogels are formed in vitro by combining fibrinogen and thrombin in an aqueous buffer. Thrombin is a serine protease, which activates fibrinogen by cleaving fibrinopeptides from the central domain of fibrinogen, thereby exposing peptide sequences known as fibrin knobs. Fibrin knobs bind corresponding fibrin holes on adjacent fibrin(ogen) molecules, which leads to self-assembly of a three-dimensional matrix. Fibrin gels can be formed in a variety of formats, depending on the specific analyses to be performed. In subsequent sections, we describe the methodology for forming fibrin gels for analysis of (1) structural properties through confocal microscopy, (2) polymerization dynamics through absorbance-based assays, (3) degradation dynamics through platelet-based assays, and (4) degradation dynamics through microscopy-based assays. In general, these protocols entail combining fibrinogen with fibrin-modifying materials in a stock buffer solution, followed by the addition of thrombin.

3.1 Confocal Microscopy

Confocal microscopy is used to analyze fibrin clot structure. In these experiments, a thin clot is prepared between a glass slide and a coverslip. These experiments only require a small clot volume (25–50 μL).

1. Prepare glass slide by first cleaning slides with 70% ethanol and Kimwipes.

2. Ensure that slides are dry. Next, place two strips of double-sided tape with enough space to allow placement of the fibrin clot. The tape will act as a spacer to provide room for the three-dimensional fibrin clot between the coverslip and glass slide (Fig. 1).

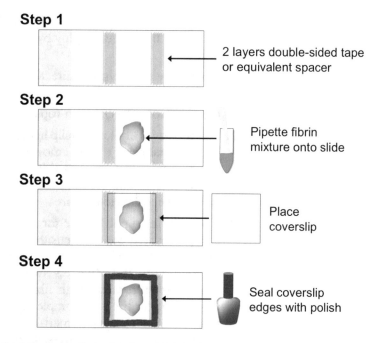

Fig. 1 Preparing fibrin clots for confocal microscopy analysis of fibrin microarchitecture. The step-by-step illustration details the setup procedure for fibrin polymerization on a glass microscope slide for characterization using confocal microscopy. A clean glass slide is prepared by layering two pieces of double-sided tape onto the slide (**Step 1**). In a microcentrifuge tube, thrombin is added to a fibrinogen solution, mixed well, and transferred onto the glass slide (**Step 2**). The fibrin clot is covered by gently placing a glass coverslip over the clot (**Step 3**) and the edges of the coverslip are sealed using nail polish (**Step 4**)

3. Next, prepare fibrinogen solution. This entails combining fibrinogen, labeled fibrinogen, 10× HEPES buffer, water, and any additional reagents (not including thrombin) into a microcentrifuge tube. 10% of the total fibrinogen concentration should be comprised of fluorescently labeled fibrinogen. As an example, if a 1 mg/mL fibrinogen clot is being prepared, 0.9 mg/mL will be comprised of unlabeled fibrinogen and 0.1 mg/mL will be comprised of fluorescently labeled fibrinogen. The appropriate amount of each stock solution is added to the tube, and the final appropriate volume is obtained by adding water. We typically utilize a 9:1 fibrinogen solution-to-thrombin solution ratio; therefore, if a 30 μL clot is being produced, the fibrinogen, buffer, and additional additive mixture will have a volume of 27 μL.

4. Prepare a 10× thrombin stock solution in water. If a final thrombin concentration of 1 U/mL is desired, a 10 U/mL stock solution should be prepared. Add the thrombin stock solution to the microcentrifuge tube containing the fibrinogen mix. If a 30 μL clot is being produced, 3 μL of the thrombin stock solution should be added to the 27 μL fibrinogen solution. Upon placing the thrombin in the tube, all components should be mixed thoroughly by pipetting up and down several times.

5. Pipet the entire solution directly onto the prepared glass slide between the strips of tape (Fig. 1).

6. Gently place a coverslip on top of the strips of tape (Fig. 1).

7. Seal the edges of the coverslip in place using nail polish (Fig. 1). To seal, first place a small amount of nail polish on two opposite corners. Next, apply a thin coating of polish around the entire slide, taking care not to use excess polish, as the polish will obstruct the viewing area.

8. Allow fibrin to polymerize for an hour prior to imaging with confocal microscope to ensure complete polymerization of fibrin network. Polymerization times vary between experimental conditions, and while less time is sufficient in certain conditions (high thrombin, high fibrinogen) lower concentrations require longer times to completely polymerize. Imaging samples before complete polymerization does not allow for appropriate comparison of steady-state network properties between different experimental conditions. However, if dynamic polymerization data is of interest, polymerization can alternatively be evaluated in real time by imaging sample at defined time intervals.

9. Image 10 μm thick sections of fibrin gel with 0.5 μm step sizes (21 sections per z-stack). Image thickness can be varied, depending on the experimental needs. The resulting images

can be viewed individually or in a 3D projection. Images are analyzed to determine various parameters such as branching, fiber number, and branch point density with image analysis tools such as ImageJ.

3.2 Absorbance-Based Polymerization Dynamics Assay

Fibrin polymerization dynamics are characterized through plate-based assays in which absorbance is monitored. In these assays, immediately following addition of thrombin, clot turbidity is measured by absorbance readings over time. As the fibrin clot begins to form, turbidity increases, ultimately reaching a steady-state turbidity upon completion of polymerization. An example of a typical polymerization curve is shown in Fig. 2a.

1. As described for the confocal microscopy protocol above, we will first prepare the fibrinogen solution in a microcentrifuge tube, again using a 9:1 volume ratio of fibrinogen solution to thrombin solutions. 80–100 µL gels can be created. The minimum recommended volume per well is 80 µL. If using 100 µL samples/well, 90 µL of the total volume will be comprised of the prepared fibrinogen solution and 10 µL will be comprised of the thrombin solution. It is recommended that each sample be run in triplicate; therefore 270 µL of fibrinogen solution is required for each experimental group. Furthermore, to account for potential losses in pipetting, it is also recommended that 10% of additional volume is prepared. Therefore, for a triplicate polymerization experiment utilizing 100 µL clot volumes, at least 297 µL total volume should be prepared. Samples should be prepared by combining fibrinogen, 10× HEPES buffer, water, and any additional reagents (not including thrombin) into a microcentrifuge tube. The appropriate amount of each stock solution is added to the tube, and the final appropriate volume is obtained by adding water.

2. Prepare a 10× thrombin stock solution in water. If a final thrombin concentration of 1 U/mL is desired, a 10 U/mL stock solution should be prepared. If 100 µL clots are being prepared using a 9:1 fibrinogen solution-to-thrombin solution ratio, 10 µL of 10× thrombin solution will be required per well. As with fibrinogen solutions, it is recommended that 10% of additional volume is prepared.

3. Pipette prepared fibrinogen solution (containing any additional additives, such as fibrin-modifying material, but not including thrombin) into all desired wells in a 96-well plate. If preparing 100 µL clots, 90 µL of fibrinogen solutions will be added to the well. Recommended maximum number of wells is 24 because reaction is time sensitive.

4. Create and test a plate reader protocol to measure absorbance at 350 nm at a read interval of 30 s–1 min for total assay duration

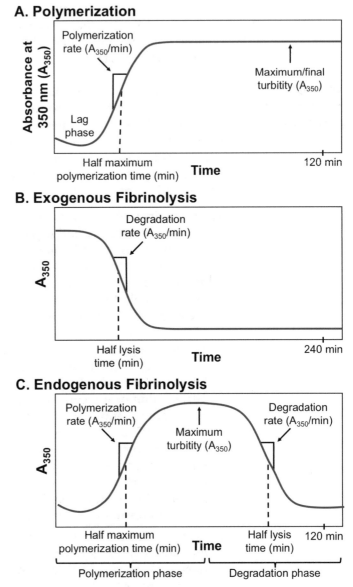

Fig. 2 Representative curves for absorbance-based polymerization and fibrinolysis assays. Fibrin polymerization and fibrinolysis dynamics can be determined by measuring absorbance values at 350 nm during assays described above. Polymerization yields a typical curve, shown in (**a**), that can be analyzed to determine various parameters including final turbidity, half-max polymerization time (time required to reach half of the maximum turbidity), and rate of polymerization. Analyzing clot degradation curves (**b**) from an exogenous fibrinolysis assay can determine clot lysis time (time at which turbidity decreases to half the maximum absorbance) and degradation rate. The endogenous fibrinolysis assay (**c**) is comprised of both a polymerization and degradation phase. The polymerization curve portion can be analyzed to determine half-max polymerization time, rate of polymerization, and maximum absorbance. The degradation portion of the curve can be used to calculate half-lysis time and rate of degradation

of 1–2 h. Ensure that the plate reader can read all the selected wells within the read interval. Most plate reader software programs do this automatically and "validate" the protocol to determine if the cumulative well read time is greater than the read interval.

5. Prior to adding the thrombin solution to the wells, measure baseline absorbance values on the plate reader at an absorbance of 350 nm.

6. Next, add 10× stock thrombin solution to wells (10 μL/well if using 100 μL clot volume and 9:1 fibrinogen-to-thrombin volume ratios). This is most easily accomplished by placing thrombin solution into a trough and using a multi-well pipette to add thrombin to the appropriate wells. Mix thrombin and fibrinogen solutions well by pipetting up and down a few times, taking care to avoid bubbles. Use fresh pipette tips to add thrombin to wells to avoid sample contamination. Wells are mixed during thrombin addition via repeated pipetting. Mixing should be performed quickly, as network polymerization starts immediately.

7. Immediately measure absorbance on plate reader according to plate reader protocol established in Subheading 3.2.4, **step 4** to monitor real-time polymerization of clot turbidity.

8. Following completion of the plate reader protocol, remove plate and save for clottability assays. Percent clottability can be determined by measuring the total protein in the clot liquor (soluble portion of the sample) remaining at the end of the polymerization assay. For clottability assays, remove 5 μL of clot liquor from each well and place into a new 96-well plate. Total protein contained in the clot liquor can then be determined using any standard protein quantitation assay, such as the Quant-it protein assay (Invitrogen) according to the manufacturer's specifications. Percent clottable protein is then calculated as [(initial soluble protein) − (soluble protein in the clot liquor)]/(initial soluble protein) × 100.

9. To analyze curves collected from the polymerization assay, subtract baseline absorbance values from each reading.

10. Analyze polymerization curves to determine various parameters including final turbidity (turbidity when polymerization is complete), half-max polymerization time (time when polymerization is half way to completion), and rate of polymerization (describes polymerization kinetics). Final turbidity is the absorbance value when polymerization is complete. This value correlates to fibrin network structure [23]. Half-max polymerization time corresponds to the time required to reach half of the maximum turbidity value. The half-max polymerization

time and rate of polymerization provide information about the rate of conversion of fibrinogen to fibrin. These parameters are shown in Fig. 2a.

3.3 Plate-Based Fibrinolysis (Degradation) Assay

Fibrinolysis is an important parameter to consider for fibrin constructs as it will determine the rates of scaffold degradation. Fibrinolysis will mediate additional cell infiltration; however if it occurs too rapidly, cell migration will be limited; therefore, it is an important process to consider. Fibrinolysis can be evaluated using two separate absorbance-based plasmin degradation assays [1], exogenous fibrinolysis via plasmin overlay and [2] tissue plasminogen activator (tPA)-induced fibrinolysis (endogenous fibrinolysis). Plasmin is a serine protease that cleaves fibrin, yielding a number of fibrin degradation products. During coagulation, plasminogen, the inactive precursor to plasmin, binds the alpha chain of fibrin and is incorporated into the developing clot. tPA, which converts plasminogen into plasmin, also binds fibrin during coagulation. Binding of both tPA and plasminogen to the polymerizing clot is required for plasminogen activation by tPA. In vivo, fibrin polymerization and degradation occur simultaneously and the initial balance of these two processes determines whether a stable clot will form. The exogenous fibrinolysis assay evaluates degradation of a preformed clot (as described above for polymerization assay) and then overlaying clots with plasmin. As the clot is degraded, turbidity is decreased, which can be monitored by measuring absorbance. The second assay described below is the tPA/plasminogen-mediated, endogenous fibrinolysis assay. In this assay, clots are prepared for polymerization in the presence of tPA and plasminogen. Following thrombin-initiated polymerization of the fibrin clot, tPA coverts plasminogen into plasmin, thereby mediating fibrin degradation. In this endogenous fibrinolysis assay, polymerization and degradation dynamics can be monitored simultaneously.

3.3.1 Exogenous Fibrinolysis via Plasmin Overlay

Prepare fibrin clots exactly as described for polymerization assay described above.

1. Prepare an equal volume of plasmin as utilized for the polymerization assay. If 100 μL of clots was prepared for the polymerization assays, 100 μL of plasmin should be prepared for each well. Prepare plasmin to have a final concentration of 0.5 U/mL and a final buffer composition of 25 mM HEPES, 150 mM NaCl, and 5 mM CaCl$_2$ pH 7.4. As in polymerization assay, it is recommended that 10% of additional volume be prepared to account for any pipetting losses.

2. Following 1–2 h of thrombin-mediated polymerization, 100 μL of plasmin is overlaid on top of the fibrin clots and gently agitated at room temperature for 4 h on a plate shaker at 100 rpm.

3. Every 30 min, record turbidity measurements (A = 350 nm) using a plate reader. At each time point, remove 5 μL samples from the clot liquor.

4. Clot liquor samples are used for subsequent determination of protein concentrations at each time point using a total protein quantitation assay, such as the Quant-It protein assay. Percent soluble protein can be reported by comparing to the initial soluble protein (prior to thrombin-initiated polymerization): 1 − [[(initial soluble protein) − (soluble protein in the clot liquor)]/(initial soluble protein)]. As the clot degrades, the percent soluble protein increases.

5. To analyze curves collected from absorbance readings, subtract baseline absorbance values from each reading. As the clot degrades, the absorbance readings will decrease over time. The resulting curves can be used to determine the rate of fibrinolysis (Fig. 2b).

3.3.2 Endogenous Fibrinolysis via Tissue Plasminogen Activator (tPA)

1. Prepare fibrin clots exactly as described for polymerization assay described above with the inclusion of human plasminogen and tPA. Clots should be prepared to obtain a final concentration of human plasminogen at 10.8 μg/mL and tPA at 0.29 μg/mL. These concentrations have been used previously and typically result in degradation within 1 h [27, 31].

2. All other steps are identical to those described in polymerization assay. Readings should be taken for 2 h to ensure completion of degradation phase of the experiment.

3. To analyze curves collected from absorbance readings, subtract baseline absorbance values from each reading. Post-assay analysis of the polymerization portion of the curves can include reporting half-max polymerization time (time required to reach half of the maximum turbidity), rate of polymerization, and maximum absorbance. The degradation portion of the curves can include calculating half-lysis time (time at which turbidity decreases to half the maximum absorbance) and rate of degradation (Fig. 2c).

3.4 Microscopy-Based Exogenous Fibrinolysis Assay

In addition to absorbance-based fibrinolysis assays, exogenous fibrinolysis can be monitored visually through a microscopy-based approach [3]. In this assay, fibrin clots are formed in a polydimethylsiloxane (PDMS) device using fluorescently labeled fibrinogen to allow for visualization. Following polymerization, plasmin is added to the clot front and clot degradation is monitored visually over time. This assay is an alternative to the absorbance-based assays if the fibrin-modifying agent significantly interferes with absorbance readings.

Prepare all solutions in the 25 mM HEPES, 150 mM NaCl, and 5 mM CaCl$_2$ buffer unless otherwise specified.

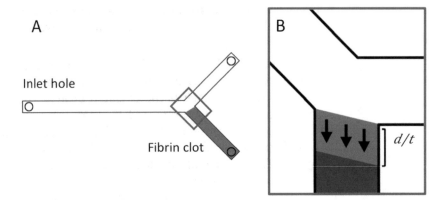

Fig. 3 Microscopy-based exogenous fibrinolysis assay setup. Three inlet holes are punched at the base of each branch in the device (**a**). The clot is pipetted in through a short branch until reaching the intersection (purple). Fibrinolytic solution is pipetted through one of the unused inlets to fill the remaining channel volume. Degradation of the clot is measured in vertical slices perpendicular to the clot boundary as it shifts and is divided by total assay time to calculate a rate of fibrinolysis (**b**)

1. Thoroughly mix Sylgard 184 in a 9:1 ratio of elastomer base to cross-linker in a final volume appropriate for the mold and degas under vacuum until no bubbles remain in solution.

2. Pour PDMS into mold and cure (*see* **Note 5**). After curing, remove device and use a biopsy punch to create inlet holes where appropriate (Fig. 3a) and store in a dust-free environment for later plasma cleaning.

3. Place clean microscope slide and device into the plasma cleaner, making sure that the two surfaces to be bound are exposed to the chamber. Clean under plasma for 1 min. After cleaning, immediately apply the device to the microscope slide in such a way that the two cleaned faces of both components are being pressed together (*see* **Note 6**). Store in a dust-free environment for at least 1 day before use.

4. Prepare clot solution in two parts: fibrinogen solution and thrombin solution. Prepare fibrinogen solution to achieve a final concentration of 2.5 mg/mL, including 10% of labeled fibrinogen. Prepare a thrombin solution to achieve a final concentration of 0.5 U/mL in the total volume of both solutions. The total volume will likely be more than what is required to fill the device adequately.

5. Mix the thrombin and fibrinogen solutions and immediately pipet the clot into the appropriate inlet of the device (Fig. 3a) until the solution reaches the "T" or "Y" intersection (*see* **Note 7**). Wait for 5 min and gently seal the inlet port with tape or grease. Allow to polymerize fully for 1 h.

6. Prepare plasmin solution to a volume appropriate to fill the remaining channel of the device at a final concentration

0.01 mg/mL. Add the plasmin solution and ensure that the channel is filled and no bubbles exist at the interface between fibrin clot and plasmin solution. Seal the inlets.

7. Acquire the boundary under the microscope and begin imaging at 5-min intervals for 12 h (*see* **Note 8**).

8. Image analysis can be performed manually or can be automated; the underlying process is identical. Adjust brightness/contrast, rotate to a workable orientation, binarize the image, and track 3–5 points along the degradation "front" as they move during the time lapse (Fig. 3b). Degradation rate is described as the distance the clot front moves divided by 12 h (*see* **Note 9**).

4   Notes

1. Human thrombin is specified in these experiments; however, thrombin derived from other species may be acceptable.

2. Typical final fibrinogen concentrations used by the authors range from 1 to 5 mg/mL; typical thrombin concentrations range from 0.25 to 2 U/mL.

3. When using higher thrombin concentrations, the final fibrin clot may not be homogenous due to local polymerization. If concentrations greater than 10 U/mL are desired, the protocol can be modified to overlay fibrinogen solution onto the thrombin solution.

4. Confocal microscopy is not accurate for measuring fiber diameter; scanning electron microscopy (SEM) can be used as an alternative method to determine fiber diameter.

5. Curing methods differ based on the mold—aluminum or other metal molds can be cured at 80° for 2 h, but silicon wafer molds are more fragile and should be allowed to cure at room temperature for 48 h.

6. Cleaning times vary based on machine and cleaning protocol. If experimental design calls for coating the channels of the device (e.g., fibrinogen coating), treat at this step when the PDMS and glass surfaces are maximally hydrophilic and amenable to reaction chemistry.

7. This step is by far the most difficult to perform. Depending on the dimensions of the device and inlet holes, the required volume of clot may be <2 μL. A steady hand and patience are essential. Larger volumes can be added via syringe and tubing, precluding the need for other sealing methods.

8. The low concentration of plasmin results in a long imaging duration. Increased concentrations of plasmin/decreased concentrations of thrombin will result in a shorter degradation time.

9. This time may be adjusted if time of assay differs from 12 h.

Acknowledgments

Funding for this work was supported by the American Heart Association (16SDG29870005), North Carolina State and the University of North Carolina at Chapel Hill.

References

1. Senior RM, Skogen WF, Griffin GL, Wilner GD (1986) Effects of fibrinogen derivatives upon the inflammatory response. Studies with human fibrinopeptide B. J Clin Invest 77:1014–1019

2. Weisel JW, Litvinov RI (2013) Mechanisms of fibrin polymerization and clinical implications. Blood 121:1712–1719

3. Brown AC, Hannan RH, Timmins LH, Fernandez JD, Barker TH, Guzzetta NA (2016) Fibrin network changes in neonates after cardiopulmonary bypass. Anesthesiology 124:1021–1031

4. Brown AC, Barker TH (2014) Fibrin-based biomaterials: modulation of macroscopic properties through rational design at the molecular level. Acta Biomater 10:1502–1514

5. Wang H, Workman G, Chen S, Barker TH, Ratner BD, Sage EH, Jiang S (2006) Secreted protein acidic and rich in cysteine (SPARC/osteonectin/BM-40) binds to fibrinogen fragments D and E, but not to native fibrinogen. Matrix Biol 25:20–26

6. Weisel JW (2005) Fibrinogen and fibrin. Adv Protein Chem 70:247–299

7. Spotnitz WD (2010) Fibrin sealant: past, present, and future: a brief review. World J Surg 34:632–634

8. Spotnitz WD, Burks S (2008) Hemostats, sealants, and adhesives: components of the surgical toolbox. Transfusion 48:1502–1516

9. Spotnitz WD, Burks S (2010) State-of-the-art review: hemostats, sealants, and adhesives II: update as well as how and when to use the components of the surgical toolbox. Clin Appl Thromb Hemost 16:497–514

10. Vaiman M, Krakovski D, Gavriel H (2006) Fibrin sealant reduces pain after tonsillectomy: prospective randomized study. Ann Otol Rhinol Laryngol 115:483–489

11. Catelas I, Sese N, Wu BM, Dunn JC, Helgerson S, Tawil B (2006) Human mesenchymal stem cell proliferation and osteogenic differentiation in fibrin gels in vitro. Tissue Eng 12:2385–2396

12. Martino MM, Mochizuki M, Rothenfluh DA, Rempel SA, Hubbell JA, Barker TH (2009) Controlling integrin specificity and stem cell differentiation in 2D and 3D environments through regulation of fibronectin domain stability. Biomaterials 30:1089–1097

13. Zhang G, Wang X, Wang Z, Zhang J, Suggs L (2006) A PEGylated fibrin patch for mesenchymal stem cell delivery. Tissue Eng 12:9–19

14. Huang NF, Lam A, Fang Q, Sievers RE, Li S, Lee RJ (2009) Bone marrow-derived mesenchymal stem cells in fibrin augment angiogenesis in the chronically infarcted myocardium. Regen Med 4:527–538

15. Takei A, Tashiro Y, Nakashima Y, Sueishi K (1995) Effects of fibrin on the angiogenesis in vitro of bovine endothelial cells in collagen gel. In Vitro Cell Dev Biol Anim 31:467–472

16. Clark RA, Nielsen LD, Welch MP, McPherson JM (1995) Collagen matrices attenuate the collagen-synthetic response of cultured fibroblasts to TGF-beta. J Cell Sci 108(Pt 3):1251–1261

17. Falvo MR, Gorkun OV, Lord ST (2010) The molecular origins of the mechanical properties of fibrin. Biophys Chem 152:15–20

18. Weisel JW (2004) The mechanical properties of fibrin for basic scientists and clinicians. Biophys Chem 112:267–276

19. Elbjeirami WM, Yonter EO, Starcher BC, West JL (2003) Enhancing mechanical properties of

tissue-engineered constructs via lysyl oxidase crosslinking activity. J Biomed Mater Res A 66:513–521

20. Girton TS, Oegema TR, Grassl ED, Isenberg BC, Tranquillo RT (2000) Mechanisms of stiffening and strengthening in media-equivalents fabricated using glycation. J Biomech Eng 122:216–223

21. Naito M, Nomura H, Iguchi A, Thompson WD, Smith EB (1998) Effect of crosslinking by factor XIIIa on the migration of vascular smooth muscle cells into fibrin gels. Thromb Res 90:111–116

22. Nair CH, Shah GA, Dhall DP (1986) Effect of temperature, pH and ionic strength and composition on fibrin network structure and its development. Thromb Res 42:809–816

23. Wolberg AS (2007) Thrombin generation and fibrin clot structure. Blood Rev 21:131–142

24. Ameer GA, Mahmood TA, Langer R (2002) A biodegradable composite scaffold for cell transplantation. J Orthop Res 20:16–19

25. Jiang B, Waller TM, Larson JC, Appel AA, Brey EM (2013) Fibrin-loaded porous poly(ethylene glycol) hydrogels as scaffold materials for vascularized tissue formation. Tissue Eng Part A 19:224–234

26. Zhao H, Ma L, Gong Y, Gao C, Shen J (2009) A polylactide/fibrin gel composite scaffold for cartilage tissue engineering: fabrication and an in vitro evaluation. J Mater Sci Mater Med 20:135–143

27. Brown AC, Stabenfeldt SE, Ahn B, Hannan RT, Dhada KS, Herman ES, Stefanelli V, Guzzetta N, Alexeev A, Lam WA, Lyon LA, Barker TH (2014) Ultrasoft microgels displaying emergent platelet-like behaviours. Nat Mater 13:1108–1114

28. Soon AS, Lee CS, Barker TH (2011) Modulation of fibrin matrix properties via knob:hole affinity interactions using peptide-PEG conjugates. Biomaterials 32:4406–4414

29. Soon AS, Stabenfeldt SE, Brown WE, Barker TH (2010) Engineering fibrin matrices: the engagement of polymerization pockets through fibrin knob technology for the delivery and retention of therapeutic proteins. Biomaterials 31:1944–1954

30. Stabenfeldt SE, Gossett JJ, Barker TH (2010) Building better fibrin knob mimics: an investigation of synthetic fibrin knob peptide structures in solution and their dynamic binding with fibrinogen/fibrin holes. Blood 116:1352–1359

31. Stabenfeldt SE, Gourley M, Krishnan L, Hoying JB, Barker TH (2012) Engineering fibrin polymers through engagement of alternative polymerization mechanisms. Biomaterials 33:535–544

32. Brown AC, Baker SR, Douglas AM, Keating M, Alvarez-Elizondo MB, Botvinick EL, Guthold M, Barker TH (2015) Molecular interference of fibrin's divalent polymerization mechanism enables modulation of multiscale material properties. Biomaterials 49:27–36

33. Chan LW, Wang X, Wei H, Pozzo LD, White NJ, Pun SH (2015) A synthetic fibrin crosslinking polymer for modulating clot properties and inducing hemostasis. Sci Transl Med 7(277):277ra229. https://doi.org/10.1126/scitranslmed.3010383

Chapter 8

Biofunctionalization of Poly(acrylamide) Gels

Julieta I. Paez, Aleeza Farrukh, Oya Ustahüseyin, and Aránzazu del Campo

Abstract

Engineering novel biomaterials that mimic closer in vivo scenarios requires the simple and quantitative incorporation of multiple instructive signals to gain a higher level of control and complexity at the cell-matrix interface. Poly(acrylamide) (PAAm) gels are very popular among biology labs as 2D model substrates with defined biochemical and mechanical properties. These gels are cost effective, easy to prepare, reproducible, and available in a wide range of stiffness. However, their functionalization with bioactive ligands (cell adhesive proteins or peptides, growth factors, etc.) in a controlled and functional fashion is not trivial; therefore reproducible and trustable protocols are needed. Amine or thiol groups are ubiquitous in natural or synthetic peptides, proteins, and dyes, and hence routinely used as handles for their immobilization on biomaterials.

We describe here the preparation of mechanically defined (0.5–100 kPa, range that approximates the stiffness of most tissues in nature), thin PAAm-based hydrogels supported on a glass substrate and covalently functionalized with amine- or thiol-containing bioligands via simple, robust, and effective protocols.

Key words Poly(acrylamide) gels, Bioconjugation, Chemoselective, Biomaterials, Mechanotransduction

1 Introduction

In vivo, cells are regulated by the coordinated presentation of biochemical and biophysical cues over multiple time and length scales [1, 2]. The cooperative effect of these multiple factors either synergistically or antagonistically is not necessarily linear and difficult to address on standard tissue culture plates, since the latter present poor control over tethered proteins and invariable mechanical properties. Engineering novel biomaterials that mimic closer in vivo scenarios requires the simple and quantitative incorporation of multiple instructive signals to gain a higher level of control and complexity at the cell-matrix interface. Such new materials will allow for rationalization of interdependencies among different

Kanika Chawla (ed.), *Biomaterials for Tissue Engineering: Methods and Protocols*, Methods in Molecular Biology, vol. 1758, https://doi.org/10.1007/978-1-4939-7741-3_8, © Springer Science+Business Media, LLC, part of Springer Nature 2018

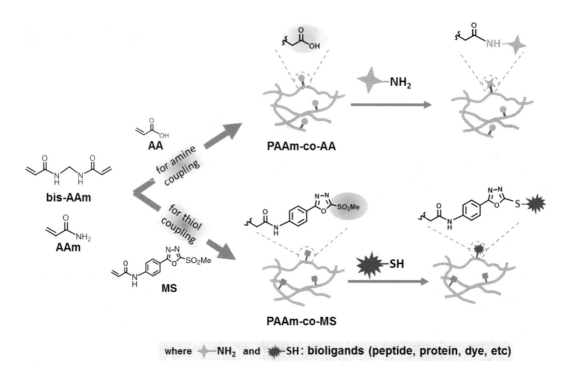

Fig. 1 Functionalization of PAAm-based gels containing specific comonomers for specific coupling of amine (top) or thiol bioligands (bottom)

cues, and will improve our understanding of the interplay between cells and their surrounding environment. With this information in hand, new approaches in tissue engineering design and regenerative medicine will become accessible [3].

Poly(acrylamide) (PAAm) gels have become very popular among biology labs as two-dimensional (2D) model substrates with defined biochemical and mechanical properties, and have been widely used in cell mechanotransduction assays [4]. These gels are cost effective, easy to prepare, reproducible, and available in a wide range of stiffness [5, 6]. Biofunctionalization of PAAm gels with bioactive ligands (cell adhesive proteins or peptides, growth factors, etc.) in a controlled and functional fashion is not trivial due to the low reactivity of the amide side groups; therefore reproducible and trustable protocols are needed.

Here, we describe the preparation of mechanically defined, thin PAAm-based hydrogels supported on a glass substrate and functionalized with amine- or thiol-containing bioligands (Fig. 1, top and bottom, respectively). Gels with stiffness values between 0.5 and 100 kPa (range that approximates the stiffness of most tissues in nature) can be prepared by adjusting comonomer and cross-linker concentrations. Once the gels are formed, amine- and thiol-containing bioligands (peptides, proteins, fluorescent dyes, etc.) at controlled surface densities can be covalently coupled via simple, robust, and effective protocols.

2   Reagents, Materials, and Equipment

All solutions are prepared at room temperature using PBS buffer. All reagents are purchased from commercial sources (analytical grade) and used as received, unless otherwise stated. The methylsulfonyl acrylamides (MS comonomers) are synthesized in our lab [7].

Reagents:

- PBS buffer (1×, pH 7.0–7.3).
- Acrylamide (AAm).
- Acrylic acid (AA).
- N,N'-methylenebis(acrylamide) (bisAAm).
- Methylsulfonyl comonomers (MS): (N-(4-(5-(methylsulfonyl)-1,3,4-oxadiazol-2-yl)phenyl)acrylamide or N-(2-(methylsulfonyl)benzo[d]thiazol-6-yl)acrylamide) [7].
- 1 M NaOH solution.
- N,N-dimethylformamide (DMF).
- 3-Acryloxypropyl-trimethoxysilane (APS).
- N,N,N',N' tetramethylethylenediamine (TEMED).
- N-(3-dimethylaminopropyl)-N'-ethylcarbodiimide hydrochloride (EDC).
- N-hydroxysuccinimide (NHS).
- 2-(N-morpho)-ethanesulfonic acid solution (MES buffer).
- NaCl.
- Pararosaniline base or other amine-containing UV-vis chromophore.
- Streptavidin and Atto-labeled biotin, or other fluorescently labeled proteins.
- Ligand of interest having either amine- or thiol-free groups for the biofunctionalization of gels.
- Magnetite nanoparticle suspension.

Other materials needed are:

- Circular borosilicate glasses (ϕ 13 mm, thickness 0.16 mm).
- Rectangular glass coverslips (76 × 26 × 1 mm).
- Plastic Petri dishes.
- Parafilm.
- Kimwipe paper towels (Kimberly-Clark, USA).
- Metallic tweezers (2aSA-SL, Erem).
- Teflon holder for coverslips.
- 1.5 mL Eppendorf vials.

- Washing bottle filled with MilliQ water.
- Washing bottle filled with ethanol.
- Nitrogen or argon gas for purging of solutions.
- Sigmacote® (a siliconizing solution for glass, Sigma Aldrich) or equivalent.
- Orbital shaker.
- Vortex mixer.
- Vacuum oven.
- Liquid nitrogen.
- Freeze-dryer.
- Vinyl-polysiloxane-based mold for hydrogel (or similar flexible material).
- Electrically conducting metal (for example, gold or platinum) for sputtering of samples.
- A removable glue for glass, like Fixogum or similar.

Equipment:

- Rheometer, Anton Paar MCR 501 or equivalent with parallel plate (25 mm) configuration.
- Scanning electron microscope (FEI-Quanta 400).
- Confocal microscope, Zeiss-LSM 880 with airy scan or equivalent.
- UV-vis spectrometer (Cary 4000) with holder appropriate for supporting coverslips. Photometric accuracy <0.00025 Abs.

Preparation of substrates and solutions.

2.1 Glass Substrates for Subsequent Acrylation

1. Use preferentially circular borosilicate glasses (*see* **Note 1**). To clean them, immerse glass substrates in ethanol and sonicate for 3 min. Rinse again with ethanol.

2. Silanization solution: In a plastic Petri dish dissolve APS (100 μL) in 95% ethanol (20 mL) (*see* **Note 2**).

2.2 Glass Coverslips for Subsequent Sigmacote Modification

Use rectangular glasses. For cleaning, immerse coverslips in ethanol and sonicate for 3 min. Rinse with ethanol and dry under nitrogen stream.

2.3 Precursor Solutions for PAAm Gel

2.3.1 Comonomer Solution for Amine Coupling (PAAm-co-AA Gels)

The exact composition of this solution depends on the targeted gel mechanical properties (*see* Table 1) [6, 8]. Dissolve AAm in PBS buffer, add AA and bisAAm, and adjust the pH of this solution to pH 8, by using 1 M NaOH solution. Bubble the comonomer solution with nitrogen or argon for 10 min to remove oxygen (*see* **Note 3**).

Table 1
Composition of PAAm-*co*-AA gels for amine coupling according to targeted gel mechanical properties

Young's modulus E (kPa)	AAm* (mg)	bis(AAm) (mg)	AA(μL)
0.5	50	0.2	5
10	120	2	12
100	180	4	18

*All quantities are for 1 mL PBS

Table 2
Composition of PAAm-*co*-MS gels for thiol coupling according to targeted gel mechanical properties

Young's modulus E (kPa)	AAm* (mg)	bis(AAm) (mg)	MS**(mg)]
0.5	50	0.2	4
10	120	2	4
100	180	4	4

*All quantities are for 1 mL PBS
**Dissolve MS in 125 μL DMF prior to use

2.3.2 Comonomer Solution for Thiol Coupling (PAAm-co-MS Gels)

Following Table 2, dissolve required amount of AAm in PBS buffer and add bisAAm [7]. Separately, dissolve MS in DMF, add it to the AAm solution, and mix well using vortex. Bubble the comonomer solution with nitrogen for 10 min to remove oxygen (*see* **Note 3**).

2.3.3 Initiator Solution

Prepare a 10%wt ammonium persulfate (APS) solution in water (dissolve 10 mg APS in 0.1 mL water) and bubble with nitrogen for 2 min (*see* **Note 4**).

2.4 Solutions Used for Gel Biofunctionalization

Prepare ligand solution according to your specific need and depending on the activity of ligand (for examples on possible concentrations, *see* **Note 5**) [7].

2.4.1 Ligand Solutions

2.4.2 EDC/NHS Solution for Activation of AA Groups

Prepare 1 mL of solution containing 38 mg of EDC (final concentration 0.2 M), and 12 mg of NHS (final concentration 0.1 M) in 0.1 M MES buffer and NaCl (0.5 M) (*see* **Note 6**) [9].

3   Methods

All procedures are performed at room temperature out of fume hood, unless otherwise stated. All employed solutions and reagents should be allowed to equilibrate at room temperature before use.

3.1   Preparation of Thin PAAm Gels

Two different kinds of modified glasses are needed for the preparation of thin gels: one is an acrylated glass and the other is a hydrophobized glass (coated with Sigmacote). The acrylated glass will support the formed hydrogel. The Sigmacote-modified coverslip is used as auxiliary to get uniform and thin gels, and is removed at the end of the preparation. During gel preparation, the polymerizing solution is sandwiched between the two slides and gel forms within, attaching only to the acrylated slide.

1. Preparation of acrylated glasses: Immerse clean substrates in silanization solution (50–70 glasses in 20 mL of silanization solution) and shake gently overnight on an orbital shaker. Remove substrates carefully with tweezers, rinse with ethanol and water, place them on a Teflon holder, and bake for 1 h at 80 °C in a vacuum oven (*see* **Note 7**).

2. Preparation of Sigmacote-modified coverslips: This step is performed in fume hood. In a plastic Petri dish, immerse 4–6 (depends on the size of dish) clean coverslips in Sigmacote (20 mL), close the container, and shake gently for 30 min. Remove substrates (*see* **Note 8**), and rinse with ethanol and water (*see* **Note 9**).

3. Preparation of PAAm hydrogel thin films: Place 200 µL of freshly degassed comonomer solution in an Eppendorf vial, add APS solution (1/100 of total volume) and TEMED (initiation catalyst, 1/1000 of total volume), and mix gently by stirring with the pipette tip without introducing air bubbles. Rapidly pipette 10 µL drops of the polymer solution onto Sigmacote-coated coverslips placed on a Petri dish (Fig. 2a) and carefully cover these droplets with acrylated glass slides (Fig. 2b) (*see* **Note 10**). Wait for 2–5 min (Fig. 2c): the hydrogel forms between the two glass slides and anchors to the acrylated side. Appearance of a rim accounts for formation of the gels (Fig. 2d). Immerse the whole assembly in water and allow swelling for 5 min (Fig. 2e). Remove the Sigmacote-modified slide by gently sliding the acrylated glass with the gel over the Sigmacote-modified one using tweezers (Fig. 2f). This step is entirely done in water. Flip the acrylate slide around and place the gels in face-up position (Fig. 2g). Keep gels in a closed Petri dish filled with water or PBS at 4 °C until further use (*see* **Note 11**).

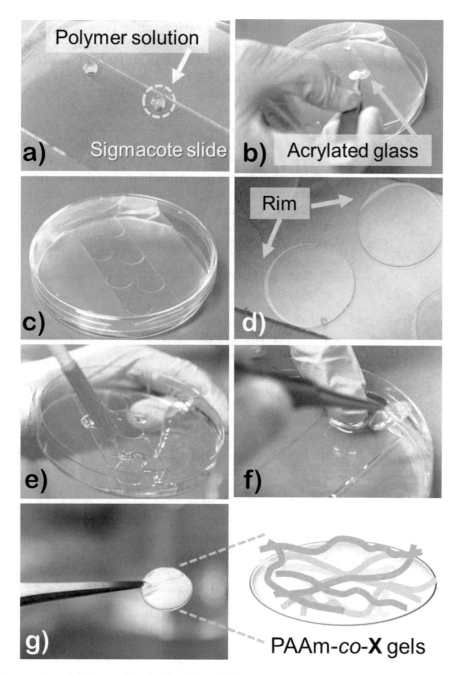

Fig. 2 Preparation of PAAm-*co*-X gels. X = AA or MS comonomer. Polymerizing solution is pipetted onto Sigmacote slide (**a**) and covered with the acrylate slide (**b**). After 5 min of resting (**c**), gel formation is indicated by the appearance of a rim in the acrylated slide (**d**). The assembly is immersed in water and the hydrogel is allowed to swell for 5 min (**e**). The Sigmacote slide is removed underwater (**f**). The formed hydrogel is anchored to the acrylated glass (**g**)

3.2 Characterization of PAAm Gels

1. Determination of gel swelling ratio: Prepare 300 µL of polymerization mixture with the desired ratio of AAm:bis(AAm) and MS or AA (*see* **Note 12**). Add initiators. Place 100 µL of gel precursor solution in a 1.5 mL Eppendorf vial and polymerize in closed vial (15 min). Prepare at least three replicas for each composition. After polymerization, remove the sample from the vial and swell the hydrogel in water for 5 min. Place samples into flasks and freeze-dry them (*see* **Note 13**). Weigh each sample ("dry weight"). Place dry hydrogels into 24-well plates filled with PBS. Weigh each sample after defined time intervals, until constant weight is achieved (generally after 12–24 h of swelling). Calculate the weight of swollen hydrogel ("wet weight"). Calculate the degree of swelling for each time interval with the following equation:

$$\text{Degree of swelling} = \frac{\text{Wet weight} - \text{Dry weight}}{\text{Dry weight}} \times 100\%$$

2. Rheologic measurement of gel stiffness (Young's modulus): Prepare molds with diameter of 20 mm and thickness of 1 mm made of vinyl-polysiloxane (*see* **Note 14**). Prepare 3 mL of comonomer solution (at least three replicates are needed) and mix with the necessary amount of initiators. On one acrylated glass place the mold, pour 330 µL of polymerizing solution (*see* **Note 15**), and cover with a Sigmacote-modified glass. Wait until the formation of rim on hydrogel. Remove carefully the flexible mold from the samples, then remove the Sigmacote-modified slide, and keep the gel in water until measurement. Place the sample into rheometer and align the plate on top. Fill water bath at the bottom with water or PBS to eliminate drying effect during measurement. Apply 0.5 N force to be sure that there is contact between hydrogel and instrument (*see* **Note 16**). Measure storage modulus (G) at 25 °C at constant strain (0.2–0.5%), and at frequency of 0.10000 rad/s. The elastic modulus (E) is calculated from the obtained storage modulus value (G), by the following equation:

$$E = 2G(1 + \nu)$$

where the Poisson number is $\nu = 0.5$.

3. Characterization of gel morphology by scanning electron microscopy (SEM): Prepare a new gel sample in bulk, i.e., without the need of a supporting acryl glass. To this end, place 100 µL of comonomer solution in a 1.5 mL Eppendorf vial, add initiator solution, and then polymerize in closed vial. Swell the obtained gel in water overnight. Freeze the hydrogel sample by carefully introducing the Eppendorf vial into liquid nitrogen (*see* **Note 17**) and then freeze-dry to remove water.

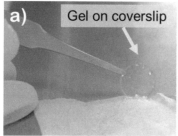  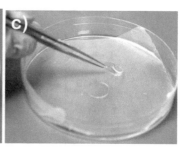

Fig. 3 Procedure for gel functionalization with ligands. The excess of water in the hydrogel is removed with a tissue paper (**a**). The ligand solution is pipetted onto a parafilm surface (**b**) and covered by the hydrogel (**c**). Ligand coupling takes place during subsequent incubation

Coat sample with an electrically conducting metal (for example, gold or platinum, 2–4 nm thickness) to inhibit charging during imaging. Image pore size and morphology with SEM at 1 kV.

3.3 Gel Functionalization with Bioligands

1. Biofunctionalization of PAAm-*co*-AA gels with amine ligands

   The carboxylic groups have to be activated before coupling of amine-containing ligands. Activate the PAAm-*co*-AA gel by immersing it in EDC/NHS solution (*see* Subheading 2.4.2) for 15 min, and then wash thoroughly with water. Remove excess water with a Kimwipe tissue (Fig. 3a) and immediately place one drop (20 μL) of the ligand solution on a parafilm surface (Fig. 3b) in a Petri dish. Overlay the drop of ligand solution with the activated gel (face down, Fig. 3c), incubate for 1 h at room temperature in closed Petri dish, and then rinse the gel 2–3 times by flushing with water for 30 s (*see* **Note 18**).

2. Biofunctionalization of PAAm-*co*-MS with thiol ligands

   Using a Kimwipe tissue, remove excess water from the hydrogel surface (Fig. 3a). Place one drop (20 μL) of the ligand solution on a parafilm surface (Fig. 3b), cover it with a PAAm-*co*-MS gel (face down, Fig. 3c), incubate for 1 h at room temperature in closed Petri dish, and then rinse the gel 2–3 times by flushing with water for 30 s (*see* **Note 19**).

3.4 Quantification of Functionalization of PAAm Gels with Bioligands

1. Measurement of film thickness and ligand distribution by confocal microscopy after modification with a fluorophore

   Functionalize PAAm-*co*-AA or PAAm-*co*-MS gels with an amine- or a thiol-containing fluorescent ligand, respectively. For example, couple the fluorophore pararosaniline base (at 0.5 or 1 mg/mL, *see* Fig. 4a, b) following the protocols in Subheadings 2.4.2 and 3.3, **step 1** (*see* **Note 20**). After functionalizing the gel with the fluorescent probe, wash the gel with water, and dry with nitrogen. Place the dry sample in the confocal microscope under 20× objective. Scan the fluorescence

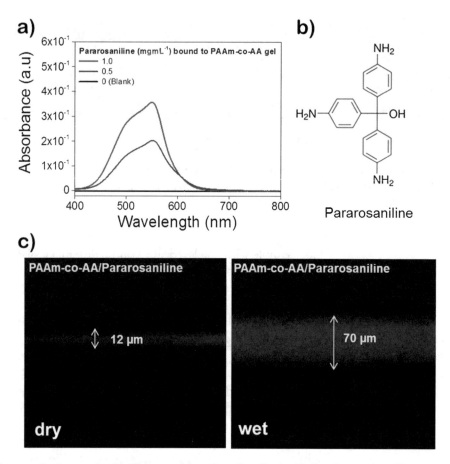

Fig. 4 One practical example of a PAAm-*co*-AA gel functionalized with the chromophore/fluorophore "pararosaniline base" (**b**). The ligand was bound to the gel in different concentrations (0.5 and 1 mg/mL), following procedures described in Subheadings 2.4.2 and 3.3, **step 1**. This bound chromophore can be tracked by UV spectroscopy (**a**) and is also useful to measure gel thickness by confocal microscopy (**c**). The dry gel thickness is 12 μm and the swollen thickness is 70 μm. Note that the fluorophore is uniformly distributed throughout the gel

signal (45% laser power) in z-axis throughout the gel until no signal is any longer observed. This sets the limits of your thickness measurement in dry state (scanning distance ca 70 μm). To measure the swollen thickness, place a drop of PBS with nanoparticles (100 μL of 1 mg/mL magnetite nanoparticle dispersion) on the surface of the gel sample and wait for equilibration. Measure the fluorescence signal along the z-axis (scanning distance ca 200 μm). The nanoparticles will allow you to identify the top layer of the gel at the microscope (*see* **Note 21**). The determined thickness in swollen state is used in the next section to calculate the ligand coupling efficiency. One example of measured thickness is shown in Fig. 4c.

2. Quantification of ligand coupling efficiency by UV-vis spectroscopy

From a given batch of prepared PAAm-*co*-AA or PAAm-*co*-MS gels, save at least 1–2 non-functionalized samples as blanks for UV measurements. Functionalize other gels from the same batch with a UV-absorbing chromophore, for example using pararosaniline base (at 0.5 or 1 mg/mL, Fig. 4a, b) following the protocols in Subheadings 2.4.2 and 3.3, **step 1**. Wash the functionalized gel with water to rinse out non-specifically adsorbed compound and dry under nitrogen stream. Using removable glue, fix the blank gel to a sample holder having a very small window and place it in the UV spectrometer. Measure absorbance twice to ensure a straight and stable baseline; scanning for 3 min is recommended. Replace the blank by the gel functionalized with the chromophore and measure absorbance (A) (*see* **Note 22**). See Fig. 4a as an example.

Calculate the ligand density (C) coupled to the gel from the measured absorbance value (A) at the chromophore's λ_{max}, by using the Beer-Lambert law:

$$C = \frac{A}{l\varepsilon}$$

where the path length l = swollen thickness of the gel (previously determined in Subheading 3.4, step 1) and εmax is the molar absorption coefficient of the chromophore at λmax (*see* **Note 23**). The binding efficiency is the ratio between initial concentration of the incubation solution and final concentration on gel.

4   Notes

1. Always prepare 20% extra glass substrates. Substrates are very thin and some break or get lost during preparation. Manipulate thin glasses with appropriate tweezers. We use homemade Teflon holders to facilitate handling of many substrates simultaneously. Glasses with other sizes and shapes can be used as well.

2. We always use freshly prepared silanization solution. Use plastic Petri dishes for the glass silanization, since undesired silanization of a glass Petri dish would result in decreased concentration of reagent.

3. The comonomer solution can be stored at 4 °C for 1 month protected from light, as bis-AAm is light sensitive.

4. APS solution (initiator) is not stable for long time. It can be stored at 4 °C for only 12 h.

5. For reference, we use the cyclic peptide cyclo[RGDfC] (0.1 mg/mL solution in PBS) or the protein fibronectin (0.01 mg/mL solution in PBS) as probes for thiol and amine coupling, respectively. Possible storage of ligand solution depends on each particular case. Ligands having UV-absorbing chromophores are useful for quantification of ligand density (*see* Subheading 3.4, **step 2**). Ligands having fluorophores are used for measurement of gel thickness and ligand distribution (*see* Subheading 3.4, **step 1**).

6. The recommended pH for EDC/NHS reaction is 5–6. Use solution immediately after preparation (it cannot be stored). 1 mL of activating solution is enough for preparation of ten samples.

7. Acrylated glass substrates prepared this way can be stored for 1 week in a well-closed container purged with nitrogen.

8. The remaining Sigmacote supernatant can be recovered and reused once or twice.

9. Substrates hydrophobized this way can be stored for 1 month in a closed container and reused several times, as long as they keep their hydrophobicity (water droplets should roll off when applied on the surface). Although other silanizing solutions (like RainX) can replace Sigmacote, we have observed that RainX coating is less stable than Sigmacote coating.

10. Make sure to appropriately space the gel precursor droplets considering the size of your acrylated slides.

11. As-prepared gels can be stored in fridge for 1–2 weeks. During this time, gels are perfectly stable for characterization purposes. For cell studies, we recommend to use freshly prepared gels, due to possible contamination of the material. If contamination happens, the gel surface will be seen as disintegrated and opaque by naked eye, with small dots on top. In that case, discard the batch and prepare a new one.

12. It is very important to consider the size of sample for swelling measurements. After freeze-drying, sample loses all water and the polymer content left will be only 5–18 wt% of the starting weight.

13. Using a coverslip to support the hydrogel during swelling, freeze-drying and weighing allows for safer handling of samples.

14. The use of a mold for gel preparation is necessary in this case due to the larger volume employed and the higher thickness attained in comparison with previous cases. Vinyl-polysiloxane is chosen because of flexibility. Any other flexible material can be chosen to prepare mold. One cylindrical side of the mold

should be open to pour monomer solution and place glass coverslip. Additionally, the diameter of the prepared sample depends on the size of your rheometer plates.

15. Using extra volume of monomer solution decreases the probability of formation of air bubbles on placing coverslips on top of monomer solution.

16. Alignment of the sample gel with the upper plate and proper adjustment of the distance between rheometer plates with slow-approaching speed may take some time. Be patient and follow the instrument indications to avoid breaking samples.

17. Always manipulate liquid nitrogen with extreme care. Wear adequate clothing and eye protection (proven for cryo-temperatures).

18. Non-specifically adsorbed ligand can be removed by washing with acetic acid solution (0.03 M). Alternatively, if the ligand contains an acid-sensitive group, repeated washing with 40% aq. ethanol can be used. To block unreacted NHS groups use aq. ethanolamine solution (10 mM) after coupling of ligand. Gels functionalized this way and immersed either in water or PBS can be stored in fridge for 1 week.

19. Unreacted MS groups can be blocked with a cysteine solution (usually 0.03 mg/mL) after coupling of ligand. Gels functionalized this way and immersed either in water or PBS can be stored in fridge for 1 week.

20. An alternative when preparing gels functionalized with fluorophores is to incubate first the gel with streptavidin (0.1 mg/mL), washing, then coupling with Atto-labeled biotin (0.5 mg/mL), and washing again. Note here that proper washing after each incubation step is crucial for minimizing background fluorescence.

21. We find this method for measuring gel thickness the most suitable one, given the highly hydrated state of the sample. To our own experience, other methods like SEM or ellipsometry have proved to be not appropriate due to the collapse of gel microstructure during drying. Moreover, fluorescence correlation spectroscopy (FCS) studies demand higher expertise in such techniques and require longer measuring times.

22. In general, low absorbance values are measured (<0.1 a.u). In case of very low absorbance (around 0.01 a.u), we recommend using the same gel before functionalization for the blank reading, in order to improve the signal-to-noise ratio.

23. For a given chromophore, we use ε_{max} value as one determined in aqueous solution.

References

1. Kyburz KA, Anseth KS (2015) Synthetic mimics of the extracellular matrix: how simple is complex enough? Ann Biomed Eng 43(3):489–500
2. Lv H et al (2015) Union is strength: matrix elasticity and microenvironmental factors codetermine stem cell differentiation fate. Cell Tissue Res 361(3):657–668
3. Banks JM, Mozdzen LC, Harley BAC, Bailey RC (2014) The combined effects of matrix stiffness and growth factor immobilization on the bioactivity and differentiation capabilities of adipose-derived stem cells. Biomaterials 35(32):8951–8959
4. Damljanović V, Lagerholm BC, Jacobson K (2005) Bulk and micropatterned conjugation of extracellular matrix proteins to characterized polyacrylamide substrates for cell mechanotransduction assays. BioTechniques 39(6):847–851
5. Denisin AK, Pruitt BL (2016) Tuning the range of polyacrylamide gel stiffness for mechanobiology applications. ACS Appl Mater Interfaces 8(34):21893–21902
6. Zouani OF, Kalisky J, Ibarboure E, Durrieu M-C (2013) Effect of BMP-2 from matrices of different stiffnesses for the modulation of stem cell fate. Biomaterials 34(9):2157–2166
7. Farrukh A, Paez JI, Salierno M, del Campo A (2016) Bioconjugating thiols to poly (acrylamide) gels for cell culture using methylsulfonyl co-monomers. Angew Chem 55(6):2092–2096
8. Moshayedi P et al (2010) Mechanosensitivity of astrocytes on optimized polyacrylamide gels analyzed by quantitative morphometry. J Phys Condens Matter 22(19):194114
9. Wirkner M et al (2011) Photoactivatable caged cyclic RGD peptide for triggering integrin binding and cell adhesion to surfaces. Chembiochem 12(17):2623–2629

Chapter 9

Synthetic PEG Hydrogel for Engineering the Environment of Ovarian Follicles

Uziel Mendez, Hong Zhou, and Ariella Shikanov

Abstract

The functional unit within the ovary is the ovarian follicle, which is also a morphological unit composed of three basic cell types: the oocyte, granulosa, and theca cells. Similar to human ovarian follicles, mouse follicles can be isolated from their ovarian environment and cultured in vitro to study folliculogenesis, or follicle development for days or weeks. Over the course of the last decade, follicle culture in a three-dimensional (3D) environment exponentially improved the outcomes of in vitro folliculogenesis. Follicle culture in 3D environments preserves follicle architecture and promotes the cross talk between cells in the follicle. Hydrogels, such as polyethylene glycol (PEG), have been used for various physiological systems for regenerative purposes because they provide a 3D environment similar to soft tissues, allow diffusion of nutrients, and can be readily modified to present biological signals, including cell adhesion ligands and proteolytic degradation facilitated by enzymes secreted by the encapsulated cells. This chapter outlines the application of PEG hydrogels to the follicle culture, including the procedures to isolate, encapsulate, and culture mouse ovarian follicles. The tunable properties of PEG hydrogels support co-encapsulation of ovarian follicles with somatic cells, which further promote follicle survival and growth in vitro through paracrine and juxtacrine interactions.

Key words Folliculogenesis, Hydrogel, Polyethylene glycol, Synthetic material, Tissue engineering, Artificial ovary

1 Introduction

In vitro culture of mouse ovarian follicles is a powerful technique to study folliculogenesis, the process of follicle development, and is a means to obtain fertilizable oocytes as a potential fertility preservation option. Ovarian follicles are the functional units of the ovary and consist of an oocyte surrounded by at least one layer of granulosa cells. The somatic compartment synthesizes and secretes hormones necessary for the orchestration of the interrelationship between other parts of the reproductive system and the central nervous system. Mammalian females are born with a finite pool of primordial follicles and their loss is irreversible. Folliculogenesis involves recruitment of primordial follicles from the resting into

Kanika Chawla (ed.), *Biomaterials for Tissue Engineering: Methods and Protocols*, Methods in Molecular Biology, vol. 1758, https://doi.org/10.1007/978-1-4939-7741-3_9, © Springer Science+Business Media, LLC, part of Springer Nature 2018

the growing pool, progressing through several developmental stages: primordial, primary, secondary, preantral multilayered secondary, and antral. The importance of cell-cell and oocyte-cell interactions in a growing follicle presents various challenges for in vitro culture in traditionally flat (two-dimensional) cultures [1]. Since the first culture of ovarian follicles was reported, the overall survival rate of the cultured follicles and maturation rates of the oocytes have steadily improved due to optimized culture media and introduction of 3-dimensional (3D) culture techniques [2]. Follicle encapsulation in a 3D environment preserves the spherical shape of the growing follicle, from the initial stages when the follicle is only 100 μm in diameter to the final stages when it reaches 350 μm in diameter [3]. Maintaining 3D architecture of a growing follicle is essential to allow for antrum formation, a fluid-filled cavity inside the growing follicle that promotes diffusion of nutrients and oxygen to the oocyte and the growing cells. Lastly, the preservation of the spherical shape maintains juxtacrine signaling between the oocyte and the surrounding granulosa through tight junctions.

Previously developed 3D culture systems have used alginate and most recently a combined fibrin-alginate system [4, 5]. These systems allowed for mouse follicle growth and maintained follicle structure but have a limited range for controlling mechanical and biological properties, such as stiffness and degradation, via chemical modification. The nondegradable nature of alginate limits the growth of follicles from large animals. For example, mouse follicles have to reach only 350 μm in diameter to be considered ready for maturation, while human follicles have to reach at least 5 mm. The growth-prohibitive properties of alginate were demonstrated recently with a human follicle that was first cultured in a 3D alginate matrix, but after it reached 500 μm in diameter the alginate was removed to allow further expansion of the follicle in a free-floating media [6]. The ex-capsulated human follicles showed improved maturation rate when first grown in alginate and then removed from the alginate and cultured in a 2D environment to allow increasing follicle expansion [6]. Therefore, a biocompatible dynamic system is needed that can be modified for properties such as cell-driven degradation and material stiffness while also accommodating the volumetric expansion of the encapsulated follicle.

We developed a 3D poly(ethylene glycol) (PEG) based system for follicle encapsulation and culture in vitro. PEG hydrogels have shown to support folliculogenesis and produce fertilizable oocytes [3]. The major advantage of PEG over natural hydrogel materials is the versatility of mechanical and biological modifications to match physiological requirements. Cross-linking PEG with peptides susceptible to degradation by different proteases, such as plasmin and MMPs, forms tunable PEG hydrogels with cell-controlled remodeling. Furthermore, PEG hydrogels can be modified with bioactive peptides such as RGD to support cell adhesion and migration [7]. Follicle co-encapsulation with support cells, such as feeder mouse embryonic fibroblasts (MEFs), allows for improved survival rate of mice follicles [8]. Most importantly, modifications of mechanical and biological properties are performed in situ in physiological conditions that are not harmful to follicles. Here, we outline a method for co-encapsulating small (100–120 μm in diameter) ovarian follicles with mouse embryonic fibroblasts within PEG hydrogels for in vitro culture. The described approach improves the survival and maturation rate of the smaller follicles that do not survive in vitro when cultured individually or without the feeder cells.

2 Materials

2.1 General Materials

1. One pair of straight fine scissors, 26 mm.
2. One pair of straight fine scissors, 24 mm.
3. One pair of straight forceps, #5.
4. One pair of curved forceps, #7.
5. Cell counter (automated or hemocytometer).
6. Dissecting microscope with a heating stage.
7. Inverted imaging microscope with imaging software.

2.2 Hydrogel Precursors

1. 8-arm poly(ethylene glycol) vinyl sulfone (PEG-VS) (molecular weight M.W. = 40000 g/mol).
2. 3-arm peptide cross-linker: YKNS (GCYKNSGCYKNSCG, M.W. = 1663.9 g/mol).
3. Cell adhesive peptide: RGD (GCGYGRGDSGP, M.W. = 1067.1 g/mol).
4. Sterile 50 mM HEPES buffer: HEPES-free acid, NaCl.
5. Sterile 1.5 mL microcentrifuge tubes.

2.3 Ovarian Follicle Isolation

1. First-generation female hybrid offsprings of two inbred strains: C57BL/6JRccHsd (maternal) CBA/JCrHsd (paternal), 12–14 days of age.
2. 70% Ethanol and dissecting mats.
3. Dissection media (DM): Leibovitz's L-15 medium, 1% heat-inactivated fetal bovine serum (FBS), and 0.5% Pen/Strep.
4. Maintenance media (MM): Minimum essential media α (αMEM), 1% FBS, and 0.5% Pen/Strep.

5. Two 35 × 10 mm sterile petri dishes.

6. One 60 × 10 mm sterile petri dishes.

7. Center well dishes for IVF: Use one (1) dish per each ovary.

8. Sterile 1.5 mL microcentrifuge tubes.

9. Two sterile insulin syringes with 27½ G needles.

2.4   Cell Culture

1. Mouse embryonic fibroblasts (MEFs).

2. DMEM culture media: DMEM, 10% FBS, 1% Pen/Strep.

3. T-75 culture flask.

2.5   Follicle and MEF Co-encapsulation and Culture

1. Microscope slides: with and without 2 mm spacers.

2. Parafilm®.

3. Sterile 96-well plates.

4. 60 × 10 mm Sterile petri dishes.

5. MM.

6. Dulbecco's phosphate-buffered saline 1× (DPBS, no calcium, no magnesium).

7. Growth media (GM): αMEM, 1 mg/mL fetuin, 3 mg/mL bovine serum albumin (BSA), 5 μg/mL insulin, 5 μg/mL transferrin, 5 ng/mL selenium, and 10 mIU/mL recombinant human follicle-stimulating hormone (rhFSH).

8. 0.25% Trypsin/EDTA.

9. Sterile 1.5 mL centrifuge tubes.

10. Sterile 15 mL conical tubes.

3   Methods (*See* Fig. 1 as an Overview)

3.1   PEG Precursor Calculation: To Be Completed the Day Before the Experiment

For follicles co-encapsulated with MEF cells, the PEG condition we use is 5% 8-arm PEG-VS modified with 0.5 mM RGD cross-linked with YKNS. Each follicle is encapsulated in 10 μL gels and based on the number of isolated follicles the total volume of the gel (V_{tot}) can be calculated. Usually, we first estimate the amounts of PEG precursors (m_{100}) in 100 μL of gel and then calculate the minimal amounts needed to reach the total final volume. Based on how much is weighed exactly ($m_{weighed}$), we adjust the volume of HEPES buffer needed to dissolve all the precursors. For all calculation, the stoichiometric ratio should be 1:1 between the VS groups available to cross-link on PEG and the functional groups (SH) on the cross-linker (YKNS).

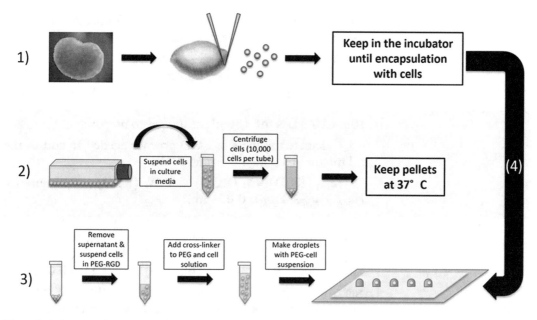

Fig. 1 General experimental outline for follicle and cell co-encapsulation; *see* text for detailed description. (*1*) The experiment should start by extracting the ovary and isolating individual follicles which are then stored in an incubator for future use up to 12 h; *see* "Ovarian Dissection" and "Follicle Isolation" steps. (*2*) Cells are then trypsinized from incubation flask, counted, and centrifuged into pellets containing 100,000 cells. Cell pellets should be maintained in DMEM at 37 °C during encapsulation for no longer than 1 h; *see* "Cell and Follicle Encapsulation." (*3*) Resuspend cell pellet in PEG-RGD solution and add cross-linker solution only when you are ready to begin encapsulation; *see* "Cell and Follicle Encapsulation." Using the PEG-RGD-cross-linker cell suspension, form droplets on Parafilm which will be used for (*4*) follicle encapsulation, as in "Cell and Follicle Encapsulation"

For the specific encapsulation condition in this chapter, 5% 8-arm PEG-VS modified with 0.5 mM RGD cross-linked with YKNS, in 100 μL gel, follow the steps below:

1. To make 100 μL of a 5% PEG gel, first calculate the mass of PEG powder needed to make that volume:

$$m_{100}\left(\text{PEG}\right)=\frac{5}{100}\times100\ \mu\text{L}\times1\frac{\text{mg}}{\mu\text{L}}=5\ \text{mg}\ ,\ \text{where}\ \frac{5}{100}\ \text{is 5\%}$$

and $1\ \dfrac{\text{mg}}{\text{mL}}$ is the assumption that the density of powder is 1.

Based on the calculated mass of PEG, then calculate the mole number of PEG (n):

$$n\left(\text{PEG}\right)=\frac{m\left(\text{PEG}\right)}{\text{M.W.}\left(\text{PEG}\right)}=\frac{5\ \text{mg}}{40,000\ \text{g}\ /\ \text{mol}}$$

$$=\frac{5}{40,000}\text{mmol}$$

$$=0.125\ \mu\text{mol}.$$

2. Calculate the mass of RGD to modify PEG with by calculating the moles of RGD:

$$n(\text{RGD}) = 0.5 \times 10^{-3} \frac{\text{mol}}{\text{L}} \times 100 \times 10^{-6} \text{L} = 0.05 \, \mu\text{mol}$$

where $0.5 \times 10^{-3} \frac{\text{mol}}{\text{L}}$ is 0.5 mM of RGD in the gel and 100×10^{-6} L is the 100 µL of the gel to prepare.

Calculate the mass of RGD powder needed to add to the gel mixture to obtain 0.5 mM RGD:

m_{100} (RGD) = n (RGD) × M.W. (RGD) = 0.05 µmol × 1067.1 = 53.4 µg ≈ 0.053 mg.

3. From here, we need to calculate the number of arms on PEG that are used to react with RGD [n (blocked)]. Since there is only 1 L-cysteine in RGD, arms of PEG blocked by RGD are equal to the mole of RGD: n (blocked) = n (RGD) = 0.05 µmol.

4. Next, we calculate the number of arms of PEG still available for cross-linking [n (available)], which is the total number of PEG arms minus the ones being used for RGD modification:

n (available) = 8 × n (PEG) − n (blocked) = 8 × 0.125 µmol − 1 × 0.05 µmol = 0.95 µmol.

5. Because all the available arms are used for cross-linking and there are three functional groups (SH from L-cysteine) on each YKNS molecule, the mole of cross-linker needed is

$$n(\text{YKNS}) = \frac{1}{3} \times n(\text{available}) = \frac{0.95}{3} \, \mu\text{mol} \cdot$$

Thus, the mass of cross-linker needed in 100 µL gel: m_{100} (YKNS) = n (YKNS) × M.W.

$$(\text{YKNS}) = \frac{0.95}{3} \times 1663.9 \, \mu\text{g} = 526.9 \, \mu\text{g} \approx 0.53 \, \text{mg}.$$

Based on how many follicles are obtained from each ovary and how many mice are used, estimate the total volume of gel (V_{tot}) required for the experiment: V_{tot} = 10 µL × the number of follicles. It is recommended to have 10% extra volume of precursors as wash and to compensate for the volume loss during mixing.

Based on the calculation for 100 µL of gel and the total volume of gel, estimate the *minimal* amounts (m_{min}) of precursors you need to weigh: $m_{\text{min}} = \dfrac{V_{\text{tot}}}{100 \, \mu L} \times m_{100}$.

In the 100 µL gel, PEG, RGD, and YKNS are mixed at a final volume ratio of V(PEG):V(RGD):V(YKNS) = 5:1:4. To adjust for the actual weighed amounts (m_{weighed}) of precursors, use the following calculation to adjust for the amount of HEPES buffer to be added to precursors (*see* **Note 1**):

$$\text{PEG}: \frac{m_{100}(\text{PEG})}{50 \, \mu L} = \frac{m_{\text{weighed}}(\text{PEG})}{m_{\text{weighed}}(\text{PEG}) + V_0(\text{PEG})}$$

$$\text{RGD} : \frac{m_{100}\left(\text{RGD}\right)}{10\ \mu\text{L}} = \frac{m_{\text{weighed}}\left(\text{RGD}\right)}{m_{\text{weighed}}\left(\text{RGD}\right) + V_0\left(\text{RGD}\right)}$$

$$\text{YKNS} : \frac{m_{100}\left(\text{YKNS}\right)}{40\ \mu\text{L}} = \frac{m_{\text{weighed}}\left(\text{YKNS}\right)}{m_{\text{weighed}}\left(\text{YKNS}\right) + V_0\left(\text{YKNS}\right)}$$

Solve for V_0 in the above equations: V_0 is the amount of sterile HEPES buffer you need to dissolve each tube of powder *prior to* mixing.

To form the gel, dissolve PEG and RGD individually first in sterile HEPES buffer, mix PEG and RGD solutions at a ratio of 5:1, and incubate at room temperature in ambient air for 15 min. Then, add YKNS to the PEG/RGD mixture to reach the final volume ratio of V(PEG):V(RGD):V(YKNS) = 5:1:4. Incubate this precursor mixture at 37 °C in ambient air and gelation should happen in 10 min.

Note that RGD and YKNS gradually lose their reactivity once dissolved in HEPES buffer. As a result, use them within *15 min* after dissolving individual components to ensure reactivity; otherwise, the gel may not form.

3.2 Ovarian Follicle Isolation

It is essential that the female mice are 12–14 days of age because at this age ovaries contain a large proportion of small follicles (100–120 µm diameter), which benefit the most from a co-culture with feeder cells [9] (Fig. 2a–c). All dissections should be performed in L15-based media (buffered for ambient air), on a 37 °C (temperature control) heated stage, and on a clean bench (laminar flow hood) to minimize potential contamination. Follicle isolation should take no longer than 30 min per ovary. For optimal results, keep the resected tissue outside the incubator for less than 1 h. To minimize pH changes, limit the exposure of αMEM-based media to ambient air. All animals are treated in accordance with the guidelines and regulations set forth by the national and institutional IACUC protocols.

1. Prior to the beginning of the experiment, sterilize all the equipment by spraying 70% ethanol and air-drying on the clean bench or in the laminar hood.

2. Prepare DM by supplementing L-15 with 1% (v/v) FBS and 0.5% (v/v) Pen/Strep. Gently invert to mix well (*see* **Note 2**). Warm up the DM to 37 °C in a water bath (*see* **Note 3**).

3. Prepare MM by supplementing αMEM with 1% (v/v) FBS and 0.5% (v/v) Pen/Strep. Gently invert to mix well (*see* **Note 4**). Prepare IVF dishes with 1 mL of MM in the center ring and 3 mL of MM in the outer ring. Utilize one IVF dish for each isolated ovary. In addition, prepare another 35 mm dish with 1 mL of MM. Pre-equilibrate all the dishes for 20 min in the incubator prior to the beginning of the isolation.

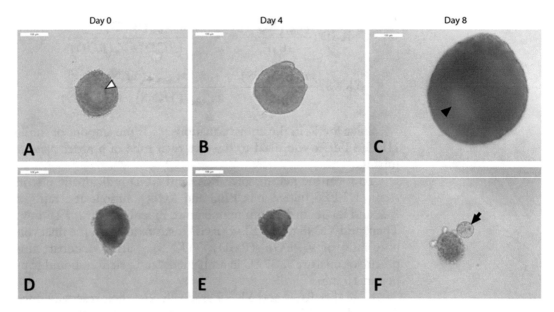

Fig. 2 Evaluation of follicle survival after the encapsulation. Top row: (**a**) a healthy follicle on day 0 has a diameter between 100 and 120 μm with centrally located round intact oocyte (white arrowhead) surrounded by granulosa cells. (**b**) A healthy follicle grows and expands and on day 8 antrum formation (**c**, black arrowhead) promotes further development. The healthy follicle will reach a diameter of 300 μm or more on day 8 or 10. Bottom row: (**d**, **e**) a non-healthy follicle has a dark complexion and the oocyte does not have a round shape with homogeneous color. Follicles with dark regions do not grow or expand and eventually extrude the oocyte (**f** black arrow points to extruded oocyte on dead follicle)

4. Euthanize one female (12–14 days of age), according to the institutional IACUC-approved protocols. Remove both ovaries from the animal. In order to assure minimal damage to the ovaries, remove parts of the oviduct and uterus around the ovary as well. Place roughly dissected ovaries into a 35 mm dish with DM.

5. Use a dissecting scope and insulin syringes to separate the ovaries from uterus, fat pad, and bursa by placing one needle at the intersection of the bursa and the oviduct to anchor the reproductive tract in place, and dissecting with the other needle. Place the second needle directly next to the first but only grip the thin membrane of the bursa. Carefully nick the bursa to expose the entire ovary. Transfer clean ovaries using curved forceps or a pipette with a blunt tip into the 35 mm dish with pre-equilibrated MM. If using forceps, try not to squish or damage the ovaries by applying excessive force. Repeat this process for both ovaries. Transfer one ovary to a 35 mm dish containing warm MM and place it in the incubator. Keep the other ovary in DM and start follicle isolation [10].

6. Set a timer for 30 min. Start isolating follicles by using two insulin syringes with 27½ G needles. With one syringe in your

nondominant hand, anchor the ovary to the bottom of the dish. And with a syringe in your dominant hand, gently tease and "flick" individual follicles from the ovary. Try to remove most surrounding ECM without puncturing the follicle. Dissect out as many follicles as possible within the 30 min. Transfer intact isolated secondary follicles (100–120 μm in diameters, 1–2 layers of granulosa cells) to the outer ring of an IVF dish with MM. When transferring, use a P10 pipette to carefully pick up isolated follicles with minimum media. To avoid follicles sticking to the walls of the same pipette tip and/or to each other, aspirate a little media first to the very end of the tip and while aspirating the follicles alternate between media and each follicle. When expelling the follicles into the outer ring of the IVF dish, use a sweeping motion such that the follicles in the tip will end up at different spots to avoid sticking to each other (*see* **Note 5**).

7. Repeat Subheading 3.2, **steps 1–6**, for the remaining ovaries to finish the isolation.

8. Select follicles for subsequent encapsulation and culture. Figure 2a shows a healthy follicle at a correct size range with a centrally located oocyte (white arrowhead) surrounded with granulosa cells. As the culture progresses the follicle grows and expands (Fig. 2b) and forms a fluid-filled cavity, antrum (Fig. 2c, black arrowhead), on day 8. Follicles that are below 100 μm in diameter and appear darker with a punctured oocyte (Fig. 2d) do not progress during culture and eventually die by extruding the oocyte (black arrow, Fig. 2f).

9. Count the number of ideal follicles for encapsulation. Ultimately, you should be able to collect between 20 and 30 good follicles from 1 ovary.

3.3 Follicle and MEF Co-encapsulation in PEG

1. Five days before the day of the experiment, seed mouse embryonic fibroblast (MEFs) cells at 8000 cells/cm² in a T-75 flask or according to established lab-specific protocols.

2. Add 10 mL of DMEM (DMEM, 10% (v/v) FBS, 1% (v/v) Pen/Strep) and change media every other day. Ensure that cells reach 80–90% confluency on the day of the experiment.

3. Prepare 50 mM HEPES buffer by dissolving 1.911 g of HEPES-free acid and 0.776 g of NaCl in 90 mL of MilliQ water. If necessary, use 1 N NaOH to titrate pH to 7.62 at room temperature such that it reaches 7.4 once warmed to 37 °C. Add water to 100 mL and measure the pH again. Sterile filter prior to use.

4. Mouse ovarian follicles are cultured in αMEM-based GM. To make GM, supplement αMEM with 10 mg/mL fetuin, 3 mg/mL BSA, 5 μg/mL insulin, 5 μg/mL transferrin, and

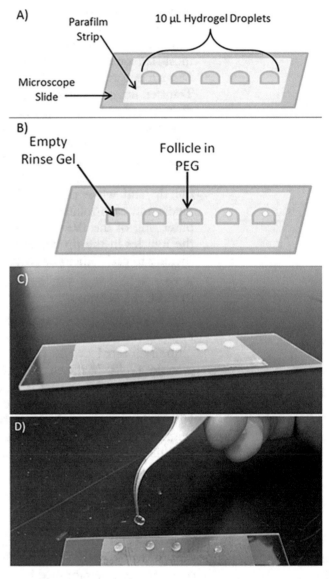

Fig. 3 Schematic of the gel preparation. (**a**) Glass slide coated with a hydrophobic layer of Parafilm. (**b**) Hydrogel droplets (10 μL each) on the hydrophobic surface for follicle rinse and follicle encapsulation. (**c**) Image of a slide with hydrogel droplets. (**d**) Image of formed hydrogel held with forceps

5 ng/mL selenium. Then, sterile filter the solution using 0.2 μm cellulose acetate syringe filter. Add 10 mIU/mL rhFSH and invert gently to mix.

5. Sterilize microscope slides and accompanying strips of Parafilm® using 70% ethanol.

6. Once dry, cover the surfaces of microscope slides with Parafilm® in a sterile environment and keep within reach for encapsulation (*see* Fig. 3a). For each workload of encapsulation

(*see* **Note 6**), you need one slide with 2 mm spacers (slide #1), one slide without (slide #2), and two accompanying Parafilm® strips to cover the working surfaces on both slides.

7. Weigh out the proper amounts of PEG, RGD, and YKNS (*see* **Note 7**). Dissolve the PEG in appropriate HEPES volume as well as RGD in its appropriate volume allowing both to become fully dissolved in the buffer.

8. In a 1.5 mL centrifuge tube, mix PEG solution with RGD solution at a 5:1 ratio and mix by pipetting up and down for 15 s. Incubate at room temperature in ambient air for 15 min.

9. In the meantime, trypsinize the cells according to established lab-specific protocols or the following. In a sterile environment, remove all the media from the flask. Briefly rinse with 2 mL of DPBS and aspirate it out. Add 3 mL of 0.25% trypsin/EDTA and incubate the cells for 3–5 min in the incubator. Check after 3 min and gently tap the flask if necessary to help MEFs detaching from the bottom of the flask.

10. After the majority (>90%) of the cells have detached from the flask, neutralize the trypsin by adding 3 mL of pre-warmed DMED culture media. Wash the surface twice by pipetting the liquid. Transfer all the liquid (~6 mL) to a 15 mL conical tube and perform cell counting.

11. Calculate the amount of cell suspension needed to encapsulate all the follicles at a concentration of 2×10^5 cells/mL (*see* **Note 8**) and aliquot into 1.5 mL microcentrifuge tubes based on how you prepare PEG, RGD, and YKNS aliquots.

12. Centrifuge at $200 \times g$ for 5 min and keep warm at 37 °C until use. Be sure to use cells within 1 h.

13. Select four good follicles for encapsulation and transfer them to the center well of an IVF dish.

14. Add the calculated amount of HEPES to dissolve YKNS and gently pipette to mix.

15. Work quickly to remove the supernatant from a 1.5 mL tube containing cells. Add YKNS to the PEG/RGD mixture to reach a final ratio of V(PEG):V(RGD):V(YKNS) = 5:1:4. Resuspend the cell pellet in appropriate amount of this precursor mixture. Start a timer for 10 min.

16. Work quickly to pipette five droplets of cell suspension in PEG precursors onto the Parafilm® on slide #1: four encapsulating droplets of 10 μL and one "wash" droplet with 20 μL of PEG solution (*see* Fig. 2a).

17. Under the dissecting scope, use a P10 pipette to transfer one of the selected follicles to the "wash" droplet first. Pipette it up and down five times in the "wash" droplet to get rid of excess media (Fig. 3b, *see* **Note 9**).

18. Change pipette tips and repeat Subheading 3.3, **step 17**, until all the four follicles are individually encapsulated in each encapsulation droplet.

19. Cover slide #1 with a Parafilm® covered slide #2. Flip the slides such that slide #1 is on top. Place the slides on a heating stage at 37 °C until gelation completes.

20. Once gelation completes, gently peel apart the slides. Using a pair of curved forceps, carefully transfer all the gels into a 60 × 10 mm sterile petri dishes containing pre-equilibrated MM.

21. Repeat Subheading 3.3, **steps 13–20**, until all the follicles are co-encapsulated with MEFs. Use new Parafilm® strips every time to ensure hydrophobicity.

22. After completing the encapsulation process, transfer the gels to 96-well plates: 1 gel per well containing 150 μL of GM.

23. Incubate at 5% CO_2 and 37 °C for 8–12 days, replacing half of the media and imaging every 2 days. On the day of the imaging, turn on the imaging microscope and set up the imaging software. Position the first 96-well plate with follicles onto the microscope. First, use a lower magnification (e.g., 5×) to find the first well with the follicle in a gel in the first row. Once you locate the hydrogel and the follicle in it in the well, switch to a higher magnification (e.g., 20×) to acquire an image of the follicle. Proceed to image all the follicles in the plate. Make sure that the imaging does not take longer than 10 min; otherwise, return the plate back to the incubator and resume imaging of the same plate after at least 30 min when GM re-equilibrates.

24. Examine the follicle morphology with a 20× objective (Fig. 2a–f). To quantitatively measure follicle growth in PEG, draw two orthogonal lines using ImageJ from basement membrane to basement membrane across the oocyte, calculate the average of the two measurements for each follicle, and generate a growth curve [2–4]. Once the follicles reach 300 μm in diameter, in vitro follicle maturation (IVFM) can be performed to evaluate oocyte quality following published procedures [2–4].

4   Notes

1. In this specific condition, the precursors are mixed at a ratio of V(PEG):V(RGD):V(YKNS) = 5:1:4. That is, for example, in a 100 μL gel, 50 μL of HEPES is used to dissolve PEG, 10 μL of HEPES is used to dissolve RGD, and the remaining 40 μL of HEPES is used to dissolve YKNS.

2. DM is buffered with HEPES for the atmospheric CO_2 level and is therefore the media used for all the procedures and manipulations performed outside the incubator. DM can be stored at 4 °C for up to 4 weeks.

3. Do not warm up DM in the incubator because the HEPES cannot buffer for CO_2 and the media becomes acidic.

4. MM is the media used to keep the isolated follicles in the incubator for all the procedures after isolation and before the transfer into GM. MM can be stored at 4 °C for up to 4 weeks.

5. Try to keep follicles separated from each other in the media at all times to avoid them sticking to each other. If follicles do aggregate, apply the same approach with the needles to pull them apart for individual follicle encapsulation.

6. Given the 10-min gelation time, after mixing all the precursors, encapsulate *no more than* four follicles. Once the user is more familiar with the process, the number of follicles in each encapsulation round can be adjusted.

7. Depending on how fast the user can encapsulate follicles and how many follicles the user encapsulates every experiment, it may be beneficial to make aliquots of PEG, RGD, and YKNS in advance, especially if less than 1 mg of the peptides is used every time to ensure accuracy. To make aliquots of PEG, RGD, and YKNS, dissolve appropriate amounts of PEG, RGD, and YKNS in sterile filtered MilliQ water with 10% of glacial acetic acid to a proper concentration, 1 mg/mL for RGD and YKNS for example. Aliquot desired amounts of each solution into tubes and freeze them at −20 °C. Snap freeze the tubes in liquid nitrogen before loading to a lyophilizer. Lyophilize until completely dry. Wrap the tubes with Parafilm® and store at −20 °C.

8. We use 2×10^5 cells/mL for MEFs and follicle co-encapsulation. The concentration of the cells is determined by the proliferation speed and nutrient consumption. MEFs are fast-proliferating cells that, if seeded at greater concentrations, can overwhelm the culture and deprive the follicle from nutrients. For slower proliferating cells, greater concentrations may be used. Careful titration is needed when co-culturing different cells.

9. It is essential to wash away the excessive media in the wash droplet to ensure successful gelation because the serum in the media interferes with and can stop the gelation from happening. If necessary, set your 10 μL pipette to 0.5 μL to avoid taking too much media when transferring.

Acknowledgments

This work is supported by Reproductive Science Program (RSP) institutional funding (NIH U046944) from the University of Michigan to AS and the National Science Foundation (NSF) Graduate Research Fellowship Program (GRFP) to UM.

References

1. Shea LD, Woodruff TK, Shikanov A (2014) Bioengineering the ovarian follicle microenvironment. Annu Rev Biomed Eng 16:29–52. https://doi.org/10.1146/annurev-bioeng-071813-105131

2. Tagler D, Makanji Y, Anderson NR et al (2013) Supplemented alphaMEM/F12-based medium enables the survival and growth of primary ovarian follicles encapsulated in alginate hydrogels. Biotechnol Bioeng 110(12):3258–3268. https://doi.org/10.1002/bit.24986

3. Shikanov A, Smith RM, Xu M et al (2011) Hydrogel network design using multifunctional macromers to coordinate tissue maturation in ovarian follicle culture. Biomaterials 32(10):2524–2531. https://doi.org/10.1016/j.biomaterials.2010.12.027

4. Shikanov A, Xu M, Woodruff TK et al (2009) Interpenetrating fibrin-alginate matrices for in vitro ovarian follicle development. Biomaterials 30(29):5476–5485. https://doi.org/10.1016/j.biomaterials.2009.06.054

5. Dolmans MM, Martinez-Madrid B, Gadisseux E et al (2007) Short-term transplantation of isolated human ovarian follicles and cortical tissue into nude mice. Reproduction 134(2):253–262. 134/2/253[pii]

6. Xiao S, Zhang J, Romero MM et al (2015) In vitro follicle growth supports human oocyte meiotic maturation. Sci Rep 5:17323. https://doi.org/10.1038/srep17323

7. Zhu J (2010) Bioactive modification of poly(ethylene glycol) hydrogels for tissue engineering. Biomaterials 31(17):4639–4656. https://doi.org/10.1016/j.biomaterials.2010.02.044

8. Tingen CM, Kiesewetter SE, Jozefik J et al (2011) A macrophage and theca cell-enriched stromal cell population influences growth and survival of immature murine follicles in vitro. Reproduction 141(6):809–820. https://doi.org/10.1530/REP-10-0483

9. Tagler D, Tu T, Smith RM, Anderson NR, Tingen CM, Woodruff TK, Shea LD (2012) Embryonic fibroblasts enable the culture of primary ovarian follicles within alginate hydrogels. Tissue Eng A 18(11–12):1229–1238

10. Xu M, Kreeger PK, Shea LD, Woodruff TK (2006) Tissue-engineered follicles produce live, fertile offspring. Tissue Eng 12(10):2739–2746

Chapter 10

Engineering Human Neural Tissue by 3D Bioprinting

Qi Gu, Eva Tomaskovic-Crook ⓘ, Gordon G. Wallace, and Jeremy M. Crook

Abstract

Bioprinting provides an opportunity to produce three-dimensional (3D) tissues for biomedical research and translational drug discovery, toxicology, and tissue replacement. Here we describe a method for fabricating human neural tissue by 3D printing human neural stem cells with a bioink, and subsequent gelation of the bioink for cell encapsulation, support, and differentiation to functional neurons and supporting neuroglia. The bioink uniquely comprises the polysaccharides alginate, water-soluble carboxymethyl-chitosan, and agarose. Importantly, the method could be adapted to fabricate neural and nonneural tissues from other cell types, with the potential to be applied for both research and clinical product development.

Key words 3D bioprinting, Bioink, Gel, Stem cells, Human neural tissue, Cell encapsulation

1 Introduction

Engineering human tissues by 3D printing cytocompatible materials with living cells enables de novo design toward customized and defined tissue fabrication. In spite of the complexity of the mammalian central nervous system (CNS), methods for bioprinting human stem cells are rapidly advancing to ensure high cell viability with efficient induction to neurons and supporting cells in constructs that in many ways recapitulate native human CNS form and function. While other 3D fabrication techniques have shown promise, including casting hydrogel scaffolds [1] with porogens that are dissolved [2] or melted [3] to regulate porosity, decellularizing native tissues to scaffolds for repopulation with cells [4], or generating organoids from self-organizing cells such as pluripotent stem cells [5], bioprinting for neural tissue engineering enables greater control over the 3D architecture of a construct, including the spatial assembly of cells and materials for optimal tissue development and fit for purpose.

Key components of any bioprinting method include a 3D model of the tissue or organ to be printed, a bioprinter operating

Kanika Chawla (ed.), *Biomaterials for Tissue Engineering: Methods and Protocols*, Methods in Molecular Biology, vol. 1758, https://doi.org/10.1007/978-1-4939-7741-3_10, © Springer Science+Business Media, LLC, part of Springer Nature 2018

under aseptic conditions, optimally viscous bioink comprising appropriate biomaterial(s), a suitable cross-linker for gelation of hydrogel-based bioinks post-printing, and cells for incorporation to the bioink or seeding to the printed scaffold. Importantly, optimization of printing extends to the rate of printing, mechanical and chemical properties of the bioink including viscosity, modulus and porosity of the scaffold, and cell density.

Here, we describe in detail a method to direct-write print frontal cortical human neural stem cells (hNSCs) for 3D neural tissue engineering. Our approach includes a novel and optimally viscous bioink that is conducive to cell survival during extrusion via a printhead and maintains its printed shape until gelation. In addition, the bioink is formulated to gel by chemical cross-linking for safe cell encapsulation, structural support, and sustained cell survival for the life of the construct. Importantly, the protocol is based on a previously published body of work relating to fabricating human neural tissue by 3D bioprinting human stem cells, which should be referred to for further details of bioink, gel, and tissue construct characterization [6]. Finally, although the method describes fabrication of neural tissue from hNSCs, it could conceivably be adapted to engineer different tissues using other stem cells.

2   Materials

2.1 Bioink and Gel Preparation

1. Alginic acid sodium salt from brown algae (alginate, Al; low viscosity, Sigma) (*see* **Note 1**).

2. Carboxymethyl chitosan (CMC; Shanghai Dibai Chemical Pty Ltd).

3. Agarose (Ag; Biochemicals).

4. 2% w/v calcium chloride.

2.2 hNSC Culture and Differentiation

1. hNSC cell line (e.g., Millipore: SCC007).

2. hNSC culture medium: NeuroCult® NS-A Basal Medium containing NeuroCult Proliferation Supplement (human; Stem Cell Technologies), supplemented with heparin (2 μg/mL; Sigma), epidermal growth factor (EGF, 20 ng/mL; Peprotech), and basic fibroblast growth factor (bFGF, 20 ng/mL; Peprotech).

3. hNSC differentiation medium: DMEM/F-12:Neurobasal Medium, 2:1 (v/v) (Life Technologies) supplemented with 2% StemPro (Life Technologies), 0.5% N2 (Life Technologies), and brain-derived neurotrophic factor (BDNF; 50 ng/mL; Peprotech).

4. Penicillin/Streptomycin (10,000 U/mL, 100×, Life Technologies).

5. Laminin (20 µg/mL; Life Technologies).

6. Digestion reagent, TrypLE Select (1×; Gibco BRL).

7. Ultra low-attachment six-well plates (Corning).

8. Standard six-well plates (Greiner Bio-One).

9. 15 mL Conical centrifuge tubes (Corning).

2.3 hNSC Bioprinting

1. 3D design software, Blender™ open-source software.

2. Stainless steel dispensing tips (Nordson).

3. 55 cc Syringe Barrels (Nordson).

2.4 hNSC Viability Assay

1. Calcein AM (5 µg/mL, Life Technologies).

2. Propidium iodide (PI, 5 µg/mL, Life Technologies).

3. Glass slides.

4. Image J software.

2.5 Cell Immunophenotyping

1. 3.7% (w/v) Paraformaldehyde (PFA; Fluka) solution in phosphate-buffered saline (PBS; pH 7.4).

2. 5% (v/v) Donkey serum (Merck) in PBS.

3. 0.3% (v/v) Triton X-100 (Sigma) in PBS.

4. Conjugated antibodies:

    (a) glial fibrillary acidic protein (GFAP; mouse, 10 µg/mL; Cell Signaling Technology).

    (b) SRY (sex-determining region Y)-box 2 (SOX2; rabbit, 10 µg/mL; Cell Signaling Technology).

    (c) vimentin (rabbit, 5 µg/mL; Cell Signaling Technology).

    (d) oligodendrocyte lineage transcription factor 2 (OLIGO2; mouse, 5 µg/mL; Millipore).

    (e) Marker of Proliferation (MKI67 (mouse, 5 µg/mL; Invitrogen).

    (f) neuron-specific class III β-tubulin (TUJ1; mouse, 10 µg/mL; Abcam).

5. Unconjugated primary antibodies:

    (a) nestin (mouse, 10 µg/mL ; Invitrogen).

    (b) synaptophysin (rabbit, 5 µg/mL; Millipore).

    (c) gamma-aminobutyric acid (GABA; rabbit, 2.5 µg/mL; Sigma).

    (d) glutamic acid decarboxylase (GAD; mouse, 2 µg/mL; Millipore).

6. 4, 6-Diamidino-2-phenylindole (DAPI, 10 µg/mL).

7. Prolong Gold antifade reagent (Invitrogen).

8. Glass coverslips and single-well/round dishes with coverslip bottom (# 1.5, 170 μm thickness).

9. Leica Application Suite AF (LAS AF) Software.

2.6 Live Cell Calcium Imaging

1. Fluo-4 (2 μM; Life Technologies).

2. Tyrode's solution (5 mM KCl, 129 mM NaCl, 2 mM $CaCl_2$, 1 mM $MgCl_2$, 30 mM d-glucose, and 25 mM HEPES, pH 7.4).

3. GABA(A) receptor antagonist bicuculline (50 μM; Sigma).

4. Leica Application Suite for Advanced Fluorescence (LAS AF) software.

2.7 Equipment

1. Magnetic hotplate stirrer.

2. 37 °C Water bath.

3. Benchtop centrifuge.

4. Humidified CO_2 incubator.

5. 3D-Bioplotter® System (EnvisionTEC GmbH).

6. Confocal microscope (Leica TCS SP5 II).

3 Methods

Reagent preparation and cell culture work should be performed in a biosafety cabinet unless otherwise specified. All media and reagents should remain sterile and pre-warmed to 37 °C. before use. Incubations and culturing should be performed in a 37 °C. incubator with a humidified atmosphere of 5% CO_2 in air.

3.1 hNSC Culture and Passaging

1. Prepare hNSC culture medium as described in Subheading 2.2.

2. Remove frozen vial of hNSCs from liquid nitrogen storage container and transfer to a water bath at 37 °C. Thaw until only one or two small ice crystals remain (*see* **Note 2**).

3. Transfer contents of thawed vial to a 15 mL Conical tube and neutralize the thawed freezing medium with 5× volume of pre-warmed culture medium, followed by centrifugation at $300 \times g$ for 3 min.

4. Discard the supernatant and seed hNSCs at a density of $2\text{-}3 \times 10^5$ cells per well of a low-attachment six-well plate with 2 mL hNSC culture medium per well.

5. During culture, the hNSCs will form into neurospheres spontaneously. Agitate daily to evenly distribute neurospheres and peform a half volume medium change after 3 days.

6. Following 5–7-day culture, collect medium containing spontaneously formed neurospheres into a 15 mL conical tube and centrifuge at $300 \times g$ for 3 min.

7. Aspirate the medium without disturbing the pellet. Optionally, the spent media can be saved for use in **step 10** below (*see* **Note 3**).

8. Add 1 mL TrypLE to the tube and gently shake the tube by hand to disperse cell pellet.

9. Place the tube in a water bath at 37 °C for 3 min.

10. Gently agitate the pellet with a pipette to separate the cells (*see* **Note 4**) and then add 5 mL hNSC culture medium (*see* **step 7** above) to stop the digestion. Centrifuge at $300 \times g$ for 5 min.

11. Aspirate the medium and resuspend the cells with fresh hNSC culture medium.

12. For bioprinting, continue to Subheading 3.2. If cells are to be further subcultured, seed the cells by repeating **step 4**.

3.2 Bioink Preparation and Bioprinting

See Fig. 1 for schematic of workflow.

1. Dissolve 1.5% (w/v) Ag in PBS in a 20 mL glass vial by heating in a microwave oven for 5 s. Allow to cool and bubbles to disperse. Repeat five to eight times.

2. Add 5% (w/v) Al and stir at 60 °C for 30 min.

3. Add 5% (w/v) CMC and stir at 60 °C for 1 h.

4. Allow the final solutions to cool to room temperature (RT) (*see* **Note 5**).

5. Collect hNSCs as described in **steps 6–10** of Subheading 3.1.

6. Suspend 5×10^6 cells in 0.5 mL bioink.

7. Load samples into a 55 cc syringe barrel and centrifuge at $300 \times g$ for 1 min to remove air bubbles.

8. Design a 3D model (e.g., 10 mm × 10 mm × 2 mm) using Blender™ software and save the file as STL type.

9. Open the STL file using Bioplotter® RP software and convert to a BPL format file.

10. Open the BPL file to establish new protocol for the 3D printing (*see* **Note 6**).

11. Set the applied pressure for bioink printing at 1.5–2.0 bar, and the temperature of the barrel and platform at 15 °C on the Bioplotter® RP software according to the manual (*see* **Note 7**). Extrusion print bioink onto a sterile glass slide in air.

12. Following printing, immerse the printed scaffolds in 2% (w/v) calcium chloride for 10 min to cross-link bioink.

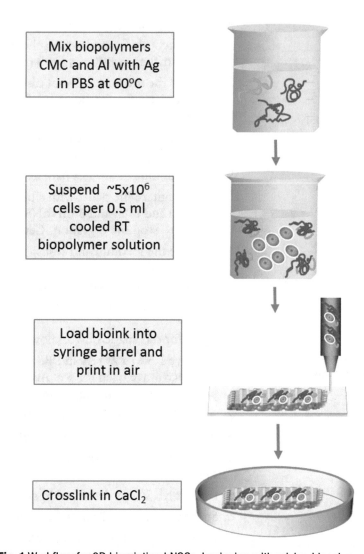

Fig. 1 Workflow for 3D bioprinting hNSCs, beginning with mixing biopolymers in PBS at 60 °C, cooling to RT for subsequent hNSC suspension, and bioink printing and gelation by ionic cross-linking

13. Immediately wash scaffold (Fig. 2) with three times 1 min washes in culture medium, followed by two 10 min washes. Incubate scaffolds for 1 h in culture medium, followed by a medium change and return to incubator for extended tissue culture (*see* **Note 8**).

3.3 In situ hNSC Culture and Differentiation

1. Incubate each construct in one well of a six-well plate with hNSC culture medium for 10 days post-printing, with half-volume medium changes performed every 2–3 days.

2. Differentiate hNSCs to functional neurons and supporting neuroglia within construct by replacing culture medium with

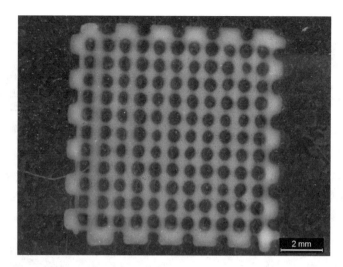

Fig. 2 Printed Al-CMC-Ag gel scaffold

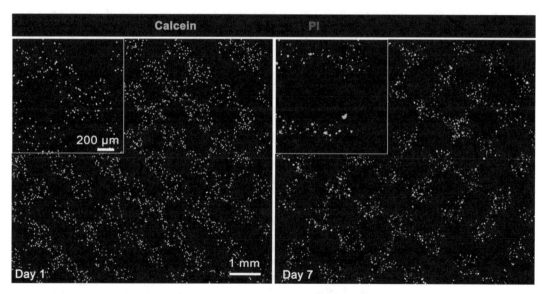

Fig. 3 Live (calcein AM) and dead (PI) cell staining of hNSCs immediately after (day 1) 3D printing and 7 days after printing, with cells visible as single cells and aggregates of cells, respectively

hNSC differentiation medium. Culture for a minimum of 14 days, including half-volume medium changes every 2–3 days.

3.4 Assessment of hNSC Viability

1. Assess hNSC viability for days 1 and 7 after printing by incubating hNSC-laden scaffolds (constructs) with calcein AM at 37 °C for 10 min followed by PI for 1 min (Fig. 3).

2. Mount the constructs onto glass slides for confocal microscopy (*see* **Note 9**).

3. Calculate the number of live and dead cells using ImageJ.

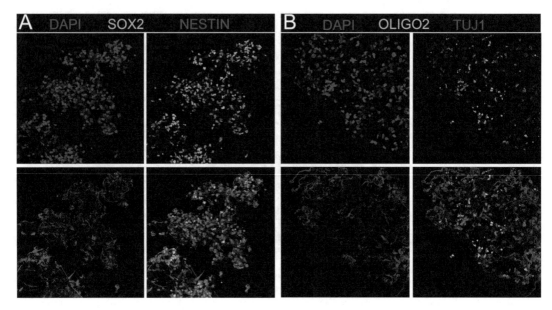

Fig. 4 (**a**) Undifferentiated printed construct 3 weeks post-printing and (**b**) differentiated hNSCs within a 3D printed gel construct 24 days post-printing including 14-day differentiation

3.5 Cell Immunophenotyping

1. Undifferentiated constructs can be immunophenotyped 3 weeks post-printing and compared to differentiated constructs 2 weeks after initiating differentiation (ie. 24 days post-printing).

2. Fix samples with 3.7% PFA at RT for 30 min, followed by concurrent blocking and permeabilization overnight at 4 °C with 5% (v/v) donkey serum in PBS containing 0.3% (v/v) Triton X-100.

3. Incubate constructs with conjugated antibodies (in the dark) or unconjugated primary antibodies at 4 °C overnight.

4. For constructs incubated with unconjugated primary antibodies, wash three times in 0.1% Triton X-100 in PBS followed by incubation for 1 h at 37 °C with species specific Alexa Fluor-conjugated secondary antibody.

5. Stain with DAPI at RT for 10 min.

6. Mount constructs with antifade reagent onto glass coverslips and image for qualitative analysis with a confocal microscope (Fig. 4). Panel A shows the neural stem marker SOX2 and NESTIN, with panel B indicating the expression of differentiation markers OLIGO2 and TUJ1.

3.6 Live Calcium Imaging

1. Incubate constructs with Fluo-4 in fresh hNSC differentiation medium for 30 min at 37 °C. Rinse the constructs in Tyrode's solution.

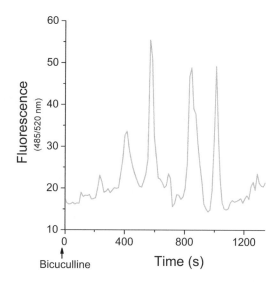

Fig. 5 Biculline-induced calcium flux of a neuron within a 3D printed construct following differentiation of hNSCs

2. Mount constructs on coverslip-bottom dishes and observe spontaneous intracellular calcium release on a confocal microscope, ideally while immersed in Tyrode's solution (*see* **Note 10**).

3. Add bicuculline to induce intracellular calcium release and immediately visualize the signal by confocal microscopy (*see* **Note 11**).

4. Images from **steps 2** to **3** above are collected and time course analysis of calcium response is performed by measuring average fluorescence intensity (Fig. 5) using LAS AF software.

4 Notes

1. While alginates have a wide range of viscosities, we routinely employ low-viscosity alginate with 100–300 cP for optimal hNSC survival during extrusion printing.

2. Use cryo-gloves to avoid exposure to skin. Using forceps, partially immerse vial in a 37 °C water bath to reduce the risk of contamination, and spray the closed vial with 70% ethanol before opening in a biosafety cabinet.

3. The aspirated medium can be collected for use in **step 10** to inactivate TrypLE.

4. Do not flux the cells by pipetting more than seven times.

5. The solutions can be stored in the refrigerator for up to 3 weeks.

6. **Steps 8–10** can be performed at any time prior to printing.

7. Before printing, it is necessary to calibrate the dispensing tip position according to the manufacturer's instructions.

8. It is preferable to use medium containing penicillin and streptomycin to prevent contamination.

9. A coverslip can be placed over the slide-mounted construct using a spacer equal to the height of the construct, being careful not to apply any pressure.

10. Tyrode's solution will assist with maintaining cell viability during imaging.

11. Bicuculline is the GABA(A) receptor antagonist that induces transient release of intracellular calcium.

Acknowledgments

The authors wish to acknowledge funding from the Australian Research Council (ARC) Centre of Excellence Scheme (CE140100012). Professor Gordon Wallace acknowledges the support of the ARC through an ARC Laureate Fellowship (FL110100196).

References

1. Gu H, Yue Z, Leong WS, Nugraha B, Tan LP (2010) Control of in vitro neural differentiation of mesenchymal stem cells in 3D macroporous, cellulosic hydrogels. Regen Med 5(2):245–253

2. Li H, Wijekoon A, Leipzig ND (2012) 3D differentiation of neural stem cells in macroporous photopolymerizable hydrogel scaffolds. PLoS One 7(11):e48824

3. Henderson TMA, Ladewig K, Haylock DN, McLean KM, O'Connor AJ (2013) Cryogels for biomedical applications. J Mater Chem B 1:2682–2695

4. De Waele J, Reekmans K, Daans J, Goossens H, Berneman Z, Ponsaerts P (2015) 3D culture of murine neural stem cells on decellularized mouse brain sections. Biomaterials 41:122–131

5. Lancaster MA, Renner M, Martin CA, Wenzel D, Bicknell LS, Hurles ME, Homfray T, Penninger JM, Jackson AP, Knoblich JA (2013) Cerebral organoids model human brain development and microcephaly. Nature 501:373–379

6. Gu Q, Tomaskovic-Crook E, Lozano R, Chen Y, Kapsa RM, Zhou Q et al (2016) Functional 3D neural mini-tissues from printed gel-based bioink and human neural stem cells. Adv Healthc Mater 5:1429–1438

Chapter 11

High-Throughput Formation of Mesenchymal Stem Cell Spheroids and Entrapment in Alginate Hydrogels

Charlotte E. Vorwald, Steve S. Ho, Jacklyn Whitehead, and J. Kent Leach

Abstract

Mesenchymal stem cells (MSCs) are a promising cell source for tissue repair and regeneration due to their multilineage capacity, potential for autologous use, and secretion of potent bioactive factors to catalyze the endogenous repair program. However, a major limitation to current cell-based tissue engineering approaches is the drastic loss of cells upon transplantation. The causation of this loss, whether due to apoptosis following a dramatic change in the microenvironment or migration away from the defect site, has yet to be determined. MSCs formed into aggregates, known as spheroids, possess a strong therapeutic advantage compared to the more commonly used dissociated cells due to their improved resistance to apoptosis and increased secretion of endogenous trophic factors. Furthermore, the use of biomaterials such as alginate hydrogels to transplant cells in situ improves cell survival, localizes payloads at the defect site, and facilitates continued instruction of cells by manipulating the biophysical properties of the biomaterial. Transplantation of MSC spheroids without a vehicle into tissue defects comprises the majority of studies to date, ceding control of spheroid function due to the cell's interaction with the native tissue extracellular matrix and abrogating the established benefits of spheroid formation. Thus, there is a significant need to consider the role of biomaterials in transplanting MSC spheroids using an appropriate carrier. In this chapter, we describe high-throughput formation of spheroids, steps for further characterization, and encapsulation in alginate hydrogels with an eye toward localizing MSC spheroids at the target site.

Key words Mesenchymal stem cell, Encapsulation, Transplantation, Spheroid, Alginate

1 Introduction

Personalized medicine remains at the forefront of clinical practice, aiming to provide treatment for individual patients based on their unique chemistries and biological makeup. The use of mesenchymal stem cells (MSCs) in cell-based approaches for tissue engineering holds great promise for improved tissue development in light of their multilineage capacity, ability to achieve large numbers of cells from culture expansion, and potential for autologous use [1]. When given the appropriate cues, MSCs can readily differentiate toward osteoblasts, chondrocytes, and adipocytes [2], yet clinically, these cells are more commonly studied for their potential to

Kanika Chawla (ed.), *Biomaterials for Tissue Engineering: Methods and Protocols*, Methods in Molecular Biology, vol. 1758, https://doi.org/10.1007/978-1-4939-7741-3_11, © Springer Science+Business Media, LLC, part of Springer Nature 2018

indirectly contribute to tissue repair through secretion of endogenous trophic factors and immunomodulatory potential [3]. However, current shortcomings include rapid loss in viability upon cell transplantation and low cell persistence at the transplant site. Forming MSCs into spheroids promotes cell-cell interaction by forced adhesion while also enabling cell-matrix interactions with endogenous cell-secreted matrix within the spheroid [4], resulting in enhanced cell survival, improved resistance to apoptosis, and increased secretion of potent trophic factors such as vascular endothelial growth factor (VEGF) [5, 6]. Spheroids have been utilized in a multitude of applications including promoting angiogenesis as a potential treatment for peripheral vascular disease [7] and chemotherapeutic screening for cancer treatment [8]. It is evident that 3D culture has great potential to mimic biological environments for therapeutic purposes.

Numerous techniques are successfully used for spheroid fabrication such as the hanging-drop method [9] and methylcellulose encapsulation [10]. Although effective at producing spherical aggregates, the hanging-drop procedure requires multiple iterations to achieve adequate cell numbers for transplantation purposes and is labor intensive. These challenges increase variability in the characteristics of resulting spheroids and limit the number of spheroids that can be produced in a given time. Methylcellulose requires additional materials and steps that yield similar results to the method described below, thereby extending the time required to form spheroids.

Here, we demonstrate that spheroids can be successfully fabricated in a high-throughput manner by rapid centrifugation in agarose molds, creating an efficient, standardized platform. This elegant design enables tailoring of spheroid size by changing the concentration of cells added to each well. The size of the spheroid is proportional to the cell concentration, based on the average number of cells per microwell. Spheroid morphology and size can be easily analyzed and quantified, further bolstering its reproducibility [9].

The therapeutic efficacy of MSC spheroids has been demonstrated in preclinical models of peripheral artery disease and wound repair [11, 12]. Compared to an equal number of dissociated cells, spheroids survived longer and were more efficient at promoting neovascularization due to increased trophic factor secretion. However, these studies are commonly performed by transplanting spheroids directly into the tissue, thereby ceding control of cell function achieved through cohesion to the surrounding extracellular matrix following adhesion to the tissue and subsequent migration [13, 14]. Substantial evidence demonstrates the efficacy of biomaterials to localize dissociated cells at the target site [15]. Therefore, one approach for sustaining the therapeutic advantages of MSC spheroids is to transplant these aggregates using hydrogel carriers that can localize and instruct cells at the implant site. There are a host of synthetic and naturally derived biomaterials that are relevant for this purpose including polyethylene glycol diacrylate (PEGDA), hyaluronic acid (HA), collagen, and fibrin [16]. As another example, alginate is a well-characterized polymer in drug delivery and tissue engineering due to its non-fouling, bioinert, and injectable properties [17, 18]. Additionally, it can be easily tailored through peptide conjugation using various chemistries to promote cell adhesion [19], permitting an understanding of signals presented independently and in combination. Compared to unmodified alginate, spheroids entrapped in arginine-glycine-aspartic acid (RGD)-modified alginate hydrogels exhibited increased viability and function [5]. Thus, encapsulation of spheroids in alginate hydrogels offers a quick, reliable mode for increased stability of spheroid function and efficacy.

A plethora of applications can be pursued with the ability to reproducibly and efficiently encapsulate spheroids of varying cell numbers and sizes in an established biomaterial. This extends beyond MSCs and could be used with varying cell types. The method described here allows for robust cell culture and will enhance research procedures for improved cell-based applications.

2 Materials

Prepare and store all reagents at room temperature unless otherwise indicated. Follow all chemical and biological waste regulations when disposing waste materials.

1. 100 mL graduated cylinder.

2. Magnetic stir plate.

3. Magnetic stir bar.

4. 100 mL glass bottle.

5. Autoclave.

6. Autoclave pouch.

7. Biological safety cabinet.

8. Weighing scale (10 mg to 220 g range).

9. 1.5% Agarose gel solution: UltraPure Agarose (Invitrogen, Carlsbad, CA).

10. Silicone rubber, 10 mm thickness (Hydrosil 1:1, SILADENT Dr. Böhme & Schöps GmbH, Goslar, Germany).

11. 100–1000 μL pipettes.

12. Sterile pipet tips.

13. Sterile tissue culture-treated culture dish (24-well plate, Falcon, Corning, NY).

14. Spatula, autoclaved at 250°C for 1 h.

15. Centrifuge with attachments that accommodate well plates.

16. Cell culture incubator.

17. Sterile tissue culture-treated culture flasks (225 cm²).

18. Cell counting device.

19. MSCs: Bone marrow-derived MSCs can be harvested and isolated as previously described [2–20]. Commercial suppliers are also available as reliable sources. Culture under standard conditions until sufficient cell numbers are achieved [2].

20. Mesenchymal stem cell culture medium: α-Modified Eagle's medium (α-MEM) supplemented with 10% fetal bovine serum, 100 units/mL penicillin, and 100 mg/mL streptomycin.

21. Microplate shaker.

22. 2.1% RGD-conjugated alginate solution (*see* **Note 1**). Store at 4°C for up to 6 months.

23. 200 mM $CaCl_2$: Store at room temperature for up to 6 months.

24. Silicone rubber, 1.5 mm thickness (3M, St. Paul, MN, USA).

25. 8 mm biopsy punch.

26. Forceps.

27. Flat dialysis membrane sheet: 3.5 kDa MWCO (Spectrum Labs, Rancho Dominguez, CA, USA). Cut as needed to specifications. *See* **Note 2**.

28. 10 cm × 10 cm glass plate: dimensions of plate can be variable. This is to provide a sterile, flat surface for the silicon mold.

29. 70% ethyl alcohol/ethanol.

30. Phosphate-buffered saline (PBS) solution: Dulbecco's phosphate-buffered saline.

31. Sterile, 50 mL reagent reservoirs.

32. Sterile 24.5 cm × 24.5 cm square culture dish (Corning, Corning, New York, USA): Dimensions of dish can be variable. This will provide a sterile surface for the glass plate.

3　Methods

Carry out all procedures in an aseptic biological safety cabinet unless otherwise specified. The following steps (Subheading 3.1, **steps 1–10**) yield one well of spheroids. Personal protective equipment (PPE) must be worn at all times.

3.1　Spheroid Formation

1. Prepare the 1.5% agarose gel solution outside of the biological safety cabinet: Measure 90 mL deionized water in a graduated cylinder and transfer to a glass bottle. Add a stir bar and place

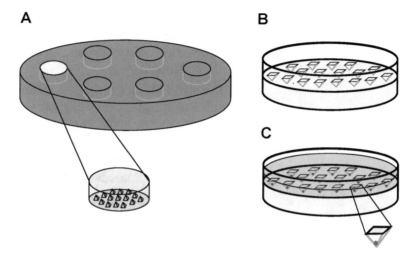

Fig. 1 Spheroid formation method. (**a**) Outline of the master mold framework. A section of PDMS is cast to produce wells with 800 μm microwell protrusions. (**b**) Acellular agarose gel in a well of a 24-well plate after centrifugation. (**c**) Agarose gel after centrifugation with cell solution. As agarose is largely non-adherent, cells aggregate at the bottom of microwells. Diagrams not drawn to scale

on a magnetic stir plate. While stirring, weigh 1.355 mg of agarose and transfer to the glass bottle and cap. Stir for 3 additional minutes. The agarose will not fully dissolve until heat is applied. Autoclave at 250 °C for 1 h, at which the agarose should be fully dissolved. *See* **Note 3**.

2. Prepare the master mold outside of the biological safety cabinet: Using 10 mm thick silicone, create a cast 6 mm in depth and 15.5 mm in diameter, with 300 microwells protruding the cast surface, each 800 μm in diameter at the base (Fig. 1a) as previously described [21]. Seal in an autoclave pouch and autoclave at 250 °C for 2 h. *See* **Note 4**.

3. Remove the master mold from the autoclaved pouch and place on a sterile flat surface in the biological safety cabinet.

4. Pipet 850 μL of 1.5% agarose into one well of the master mold. Allow to cool to room temperature for 10 min. *See* **Note 5**.

5. Remove the agarose gel from the master mold and gently place it into a 24-well plate (Fig. 1b). Use a spatula to push down the agarose gel as far as possible without disturbing the microwells. The microwells should be facing up from the plate surface. *See* **Note 6**.

6. Pipet 1 mL of media on top of each agarose mold. This will add weight to ensure that the agarose gel is properly secured to the bottom of the plate.

7. Secure the lid onto the 24-well plate. Outside of the biological safety cabinet, centrifuge the plate at 290 × *g* for 3 min. Inside the biological safety cabinet, aspirate excess media from the well. *See* **Note 7**.

8. Harvest the MSCs from tissue culture plates and quantify the number of available cells using any appropriate counting device (i.e., cell counter, hemacytometer).

9. Suspend MSCs in serum-containing complete α-MEM culture medium at the concentration for the desired number of cells for 1 mL per well. *See* **Note 8**.

10. Pipet 1 mL cell suspension into one well of the 24-well plate containing the agarose mold. Pipet up and down gently several times to evenly disperse cells (Fig. 1c). Secure the lid onto the plate. *See* **Note 9**.

11. Outside of the biological safety cabinet, centrifuge the 24-well plate at 163 × *g* for 8 min.

12. Incubate the 24-well plate in 37°C for up to 48 h. *See* **Note 10**. Fig. 2a, b illustrates the result of spheroid formation after 48 h with 5,000 or 10,000 cells/spheroid. Spheroids with 10,000 cells are visibly larger in the microwell, also indicated by a greater diameter in Fig. 2c.

3.2 Spheroid Entrapment in Alginate Hydrogels

1. Prepare the 2.1% RGD-conjugated alginate solution: In a 50 mL conical tube, add the necessary volume of sterile α-MEM to RGD-modified alginate to obtain an alginate solution of 2.1% by weight. *See* **Note 11**. Tape conical tube to a microplate shaker, and set speed to 18 × *g* and temperature to 37°C to allow alginate to mix overnight.

2. Prepare the CaCl$_2$ solution: Weigh 14.701 g of solid calcium chloride dehydrate (molecular weight 147.01 g/mol). Mix and dissolve in 500 mL deionized H$_2$O. Sterile filter into a sterile 500 mL glass bottle.

3. Prepare the silicone mold: Place 1.5 mm thick silicone rubber on sterile surface and use an 8 mm biopsy punch to punch out desired number of holes approximately 1 cm apart.

4. After spheroid formation, typically 1–2 days, collect the spheroids from the well plate by gently pipetting up and down to displace spheroids from the microwells. If needed, wash the well with extra media to collect remaining spheroids. Pipet the spheroids into a 15 mL conical tube.

5. Allow spheroids to settle to the bottom of the conical tube. Alternatively, gently spin down tube at 163 × *g* for 1 min. Carefully aspirate the media. *See* **Note 12**.

6. Resuspend spheroids in 2.1% alginate solution (Fig. 3a). The concentration of spheroids depends on the cell concentration

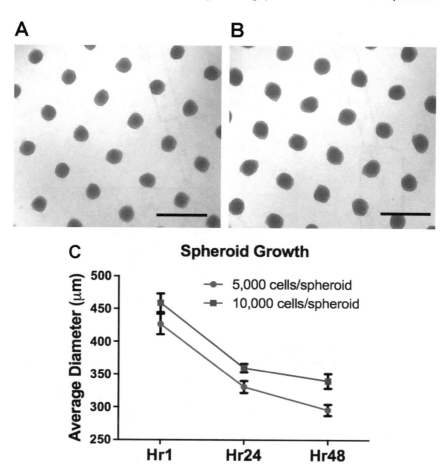

Fig. 2 Spheroid size and diameter measurement. Spheroid formation in agarose well after 48 h with (**a**) 5,000 cells/spheroid or (**b**) 10,000 cells/spheroid. (**c**) Spheroid diameter measurements for 5,000 and 10,000 cells/spheroid over time

desired in alginate hydrogels. We suggest at least 5×10^6 cells per mL to ensure adequate cellular response for necessary assays.

7. Within the biosafety cabinet, set up gelling area. Place the glass plate in a sterile, square culture dish. Place silicon mold on top of the glass plate and make sure that it is firmly adhered. *See* **Note 13**.

8. Pipet 90 µL of alginate/spheroid solution into each well in the silicone mold (Fig. 3b). *See* **Note 14**.

9. Using sterile forceps, pick up the dialysis membrane from PBS solution and dab corners on sterile gauze to remove excess liquid. Carefully place dialysis membrane over the wells of the silicone mold and use forceps to flatten completely. *See* **Note 15**.

10. Pipet 200 mM CaCl$_2$, 3–4 mL, over the membrane to cover the wells (Fig. 3). Let the solution sit for 5 min, which allows

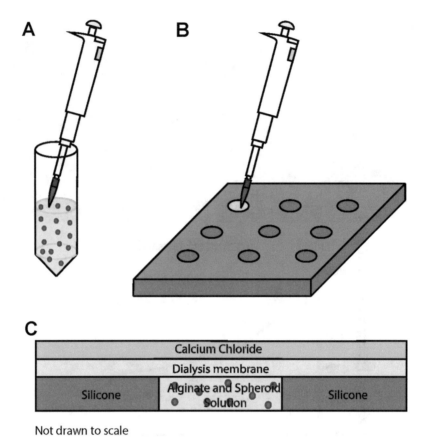

Fig. 3 Spheroid entrapment in alginate hydrogels. (**a**) Spheroids suspended in 2.1% alginate solution. (**b**) Layout of silicone mold for incorporation of spheroids in alginate suspension. (**c**) Cross-sectional view of alginate hydrogel assembly for cross-linking initiation

calcium ions to diffuse into the hydrogel discs. Carefully slide the membrane off laterally and pipet more CaCl₂ solution over the wells for direct contact and immersion. Alginate cross-linking proceeds for an additional 5 min. *See* **Note 16**.

11. After 10 min of alginate cross-linking is complete, tilt the glass plate to remove CaCl₂ solution. Use a sterile spatula to carefully remove excess alginate on top of the silicone mold.

12. Hydrogels can now be lifted and transferred to a 24-well plate for culture in the desired culture medium. These can be transplanted in vivo or maintained in culture for in vitro studies. *See* **Note 17**.

4   Notes

1. A variety of peptides can be conjugated using various chemistries. We illustrate the use of RGD due to its widespread use and ease of conjugation via carbodiimide chemistry [15–22].

2. The dialysis membrane must be cut and hydrated in deionized water for 10 min prior to use. After hydration, membranes should be placed into a plastic container containing 70% ethanol to be sterilized for 30 min. After sterilization, place membranes in a sterile reagent reservoir containing sterile PBS to rinse off residual 70% ethanol.

3. Agarose solution will solidify at room temperature. For repeated use, store at room temperature and reheat on a hot plate, using the stir bar to create a homogeneous solution. Alternatively, a microwave can be used with subsequent stirring.

4. Ensure that the thickness of the mold will allow enough room for media when the agarose gel is placed into a 24-well plate (Subheading 3.1, **step 5**). The geometry and dimensions can be customized to achieve molds that yield aggregates of other shapes.

5. The agarose gel volume should be filled enough to form a concave meniscus over the top of the well. This is to make sure that the well is not underfilled, which can allow for media to rush under the agarose and disturb spheroids during the formation process.

6. Be careful to avoid damaging the mold. Tilt the spatula and press the agarose along the side of the gel, making sure not to directly touch the microwells.

7. If the mold is still not set at the bottom of the well, repeat this step at a higher revolution speed. Avoid speeds of $3220 \times g$ because these forces can deform the agarose. Centrifuge accessories that can accommodate well plates are necessary.

8. It is important that the desired number of cells is suspended in 1 mL of media for each well. This volume allows for adequate oxygen supply and prevents the well from overflowing.

9. Avoid creating bubbles, as this will affect cell viability. Additionally, ensure that the cell density is low enough that the cells do not overflow and create poorly formed spheroids. The maximum total cell number for a microwell with 800 μm base diameter is 20,000 cells.

10. Place the well plate in a location with minimal movement. Movement of the plate during formation can disrupt the integrity of the aggregates and result in either poorly formed aggregates or inconsistent spheroid sizes.

11. A 2.1% (w/v) solution is used to account for the small amount of volume occupied by cells. When combined, the two components will form a 2% (w/v) solution of alginate and cells. Varying percentages of alginate may also be used, as well as different types of alginate.

12. Use a micropipette to gently pipet media out of the conical tube. This allows for more controlled suction. If using a vacuum line with Pasteur pipette, adjust vacuum line to allow for gentle aspiration.

13. Complete adherence of the silicone mold to the glass plate is critical. Make visual checks beneath the plate to see if there are any air pockets and from a lateral view to make sure that there are no areas where the mold is lifted off the plate. This ensures that no alginate will leak under the mold. Mold will lift off if contact is too dry. To address this, a small amount of PBS can be applied to allow for proper adherence.

14. Alginate volume should overflow and form a concave meniscus over the top of the well. This is to make sure that the well is not underfilled, which can create bubbles that displace the alginate. Excess alginate will be spread off to the side when the dialysis membrane is placed over the mold.

15. Place the dialysis membrane on top of the mold by flattening it from the bottom up. Be careful not to lift up the membrane once it has made contact with the mold. This will cause bubbles to form in the well, as the alginate sticks to the membrane.

16. Membrane must be slid off laterally. If removed vertically, it will also remove alginate from the well. Sliding to the side will retain alginate in the wells. Do not reuse membrane, as alginate may obstruct pores and prevent efficient dialysis on subsequent gels.

17. It is prudent to remove excess alginate from the mold and encircle the wells with a spatula by tracing the gel circumference before removal so as not to lift up residual material or cause rips in the hydrogel.

Acknowledgment

This work was supported by NIH Grant R01-DE025475 to JKL. The content is solely the responsibility of the authors and does not necessarily represent the official views of the National Institutes of Health. The funders had no role in the decision to publish, or preparation of the manuscript. CEV was supported by the T32 Training Program in Basic and Translational Cardiovascular Science (T32 HL086350). JW was supported by the National Science Foundation Graduate Research Fellowship (1650042). SH was supported by the T32 Animal Models of Infectious Disease Training Program Kirschstein-NRSA (T32 AI060555).

References

1. Caplan AI (2005) Mesenchymal stem cells: cell-based reconstructive therapy in orthopedics. Tissue Eng 11(7–8):1198–1211

2. Pittenger MF et al (1999) Multilineage potential of adult human mesenchymal stem cells. Science 284(5411):143–147

3. Caplan AI (2016) MSCs: the sentinel and safe-guards of injury. J Cell Physiol 231(7):1413–1416

4. Murphy KC et al (2016) Mesenchymal stem cell spheroids retain osteogenic phenotype through alpha2beta1 signaling. Stem Cells Transl Med 5(9):1229–1237

5. Ho SS et al (2016) Increased survival and function of mesenchymal stem cell spheroids entrapped in instructive alginate hydrogels. Stem Cells Transl Med 5(6):773–781

6. Bhang SH et al (2012) Transplantation of cord blood mesenchymal stem cells as spheroids enhances vascularization. J Tissue Eng Regen Med 6:295–295

7. Wenger A et al (2004) Modulation of in vitro angiogenesis in a three-dimensional spheroidal coculture model for bone tissue engineering. Tissue Eng 10(9–10):1536–1547

8. Thoma CR et al (2014) 3D cell culture systems modeling tumor growth determinants in cancer target discovery. Adv Drug Deliv Rev 69:29–41

9. Murphy KC, Fang SY, Leach JK (2014) Human mesenchymal stem cell spheroids in fibrin hydrogels exhibit improved cell survival and potential for bone healing. Cell Tissue Res 357(1):91–99

10. Hattermann K, Held-Feindt J, Mentlein R (2011) Spheroid confrontation assay: a simple method to monitor the three-dimensional migration of different cell types in vitro. Ann Anat 193(3):181–184

11. Bhang SH et al (2011) Angiogenesis in ischemic tissue produced by spheroid grafting of human adipose-derived stromal cells. Biomaterials 32(11):2734–2747

12. Ylostalo JH et al (2012) Human mesenchymal stem/stromal cells cultured as spheroids are self-activated to produce prostaglandin E2 that directs stimulated macrophages into an anti-inflammatory phenotype. Stem Cells 30(10):2283–2296

13. Lee JH, Han YS, Lee SH (2016) Long-duration three-dimensional spheroid culture promotes angiogenic activities of adipose-derived mesenchymal stem cells. Biomol Ther 24(3):260–267

14. Sart S et al (2014) Three-dimensional aggregates of mesenchymal stem cells: cellular mechanisms, biological properties, and applications. Tissue Eng Part B Rev 20(5):365–380

15. Qi CX et al (2015) Biomaterials as carrier, barrier and reactor for cell-based regenerative medicine. Protein Cell 6(9):638–653

16. Caliari SR, Burdick JA (2016) A practical guide to hydrogels for cell culture. Nat Methods 13(5):405–414

17. Augst AD, Kong HJ, Mooney DJ (2006) Alginate hydrogels as biomaterials. Macromol Biosci 6(8):623–633

18. Lee KY, Mooney DJ (2012) Alginate: properties and biomedical applications. Prog Polym Sci 37(1):106–126

19. Hersel U, Dahmen C, Kessler H (2003) RGD modified polymers: biomaterials for stimulated cell adhesion and beyond. Biomaterials 24(24):4385–4415

20. Soleimani M, Nadri S (2009) A protocol for isolation and culture of mesenchymal stem cells from mouse bone marrow. Nat Protoc 4(1):102–106

21. Dahlmann J et al (2013) The use of agarose microwells for scalable embryoid body formation and cardiac differentiation of human and murine pluripotent stem cells. Biomaterials 34(10):2463–2471

22. Rowley JA, Madlambayan G, Mooney DJ (1999) Alginate hydrogels as synthetic extracellular matrix materials. Biomaterials 20(1):45–53

Chapter 12

Crimped Electrospun Fibers for Tissue Engineering

Pen-hsiu Grace Chao

Abstract

Collagen fibers exist in many parts of the body as parallel bundles with a wavy morphology, known as crimp. This crimp structure contributes to the nonlinear mechanical properties of the tissue, such as ligament, blood vessels, and intestine, which provide elasticity and prevent injury. To recapitulate the native collagen crimp structure, we report a robust method using electrospinning and post-processing to generate parallel polymeric fibers with crimp that simulate the structure-function relationship of native tissue mechanics. In addition to recreating the mechanical functionalities, these fibers are instructive for cell morphology and phenotype and can serve as a platform to study cell-material interactions in a biomimetic physical microenvironment.

Key words Biomaterials, Electrospinning, PLLA, Tissue engineering, Collagen, Nanofibers, Biomechanics

1 Introduction

In many collagen-rich soft tissues, such as blood vessels, intestines, valve leaflets, ligaments, and tendons, parallel collagen fiber bundles are found in a periodic wavy pattern called crimp. This structure contributes to the mechanical functionality of the tissue, where gradual unfolding of the waves with increasing strain, in addition to fiber-fiber interactions, leads to increasing tissue stiffness [1]. This strain-stiffening effect provides flexibility at low strain and prevents overloading and injury with increase in extension. In an effort to recapitulate this structure-function relationship, several approaches have been made to generate materials that exhibit the nonlinear mechanical functionality. Woven/braided fibers have been successful in reproducing the mechanical functionality of ligament and tendons [2, 3]. At the cellular level, however, these large dense fiber bundles are vastly different from the structure of the native microenvironment [3]. Researchers have synthesized fibers with wavy morphologies for a variety of applications. Conjugated spinning of two parallel yet attached polymers of different shrinkage properties generates highly crimped fibers, although not with biocompatible materials and with size scales orders of magnitude larger than what is biologically relevant [4]. Micromolding of synthetic collagen fibers with a template provides a highly aligned array of microfibers that display a well-defined microcrimped pattern [5]. This approach, however, is time consuming and labor intensive, and can only achieve materials of a limited scale.

Electrospinning has been widely used to generate micro- to nano-fiber materials in tissue engineering to mimic the native fibrous microenvironment. We recently developed a robust post-processing method to induce cell-scale crimps in aligned electrospun polymeric fibers [6]. Changing the polymer crystallinity through either a weak solvent or brief heating over the glass transition temperature (Tg) leads to fiber crimping with a periodicity of approximately 100 μm, which is on the same length scale of a cell. The resulting material exhibits the same nonlinear stress-strain behaviors as ligament and tendons with an extended toe region. When fibroblasts are seeded in the microcrimped fibers, cells elongate and extend processes along the fibers and exhibit different morphology compared with those seeded on straight fibers. This morphological change results in enhanced collagen gene expression, as well as altered responses to mechanical perturbations, demonstrating the importance of recapitulating the cellular microenvironment in the design of biomaterials for tissue engineering.

2   Materials and Equipment

2.1   Materials

2.1.1   Electrospinning Materials

1. Polylactide (PLLA, e.g., Aldrich 81273).
2. 1,1,1,3,3,3-Hexafluoro-propanol (HFP, e.g., Aldrich 105228).
3. Aluminum foil.
4. Parafilm.
5. 30 mL Syringes.
6. 19-gauge blunt-end needles.
7. Silicon tubing.

2.1.2   Cell Seeding Materials

1. 35% Hydrogen peroxide.
2. Phosphate-buffered saline (PBS, e.g., Hyclone SH30256).
3. Dopamine hydrochloride (Sigma H8502).
4. Tris–hydrochloride (JT Baker 4103).
5. Fibronectin (Sigma F1141).
6. Dulbecco's modified Eagle medium (e.g., ThermoFisher Scientific 11885).

7. Fetal bovine serum (FBS, e.g., ThermoFisher Scientific 10437).

8. Antibiotic-antimycotic, 100× (e.g., ThermoFisher Scientific 15240-062).

2.2 Equipment

2.2.1 Electrospinning System

There are several commercially available electrospinning systems, some of which allow customization. In general, a high-voltage power supply and a syringe pump are necessary to generate randomly organized fibers. In order to mimic the structure of collagen fibers, we employ a rotating collector to generate parallel fibers (for a review of designs, *see* Ref. 7). Moreover, to achieve a large and uniform material, the spinneret is attached to a "fanner" that moves along the length of the collector, thus facilitating uniform distribution of the material [8]. The following section describes our current system:

1. Power supply (e.g., Gamma High Voltage Research, ES30N-5W).

2. Syringe pump (e.g., New Era Pump Systems Inc. NE-300).

3. Rotating collector (custom):

   A rotating collector is consisted of a mandrel and a motor with controllable speed (Fig. 1). Depending on the mandrel size and polymer properties, surface speed of the collector can range from 5 to 30 m/s to achieve the desired fiber alignment. One crucial design criteria is that the collecting mandrel and motor need to be well insulated to prevent the high voltage from the power supply from damaging the motor.

4. Fanner (custom, optional):

   The purpose of the fanner is to move the spinneret in order to evenly distribute the jet along the collector. The fanner is not necessary if only a small sample is desired. Several design possibilities can achieve this goal. As illustrated in Fig. 2, we

Fig. 1 Rotating collector, with the collecting mandrel to the left and motor to the right

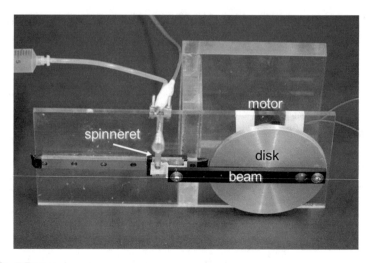

Fig. 2 Fanner

Fig. 3 The electrospinning system with the syringe pump, fanner, and rotating collector. The power supply is out of view

use a disk connected to a small motor. Rotation of the disk is translated with a beam to linear movement along a motion guide, connected to the spinneret. The fanner should also be insulated from the high voltage to prevent damage.

The full electrospinning setup as shown in Fig. 3 shows that the spinneret is attached to the fanner and connected to the power supply. The collector is connected to ground.

2.2.2 Additional Equipment

1. Orbital shaker.
2. Hot plate with digital temperature control (e.g., Thermo Scientific Nuova Hotplate).

3   Methods

3.1   Preparation of Polymer Solution

1. Seal the tip of a 30 mL syringe with parafilm.

2. Dissolve 1.7 g of PLLA in 31.92 g of HFP, reaching a final concentration of 8.5% (w/v, *see* **Note 1**).

3. Allow the solution to fully dissolve overnight on an orbital shaker at the lowest setting of the shaker.

3.2   Electrospinning of Aligned Fibrous Scaffold

1. Cover the collecting mandrel with aluminum foil using double-sided tape. Keep the foil as smooth as possible since wrinkles on the foil may lead to unevenness in the final material.

2. Mount the syringe with polymer solution on the syringe pump and set the flow rate to 2.5 mL/h (*see* **Note 2**). Connect silicon tubing between syringe and needle if using a fanner. Turn on the rotating collector.

3. Check distance between needle tip and collector to be 13 cm and the power supply is set at 13 kV to achieve a field strength of 1 kV/cm (*see* **Note 1**).

4. Confirm steady rate of flow from the needle tip and start the fanner, if using, before starting the power supply.

5. Turn on the power supply and confirm the establishment of spinning stream.

6. When the electrospinning is completed (*see* **Note 3**), turn off the power supply, syringe pump, fanner, and collector, in this order. Remove the material from the collector by cutting along the seam of the foil on the collecting drum.

3.3   Crimp Induction

1. Cut material slightly larger than the desired final size. For example, for the final dimension of 4 × 4 cm, cut to 5 × 5 cm.

2. Put material between two frosted glass plates and place the sandwich on a hot plate (preheated to 85 °C) for 15 min.

3. Remove the sandwich from the hot plate and allow it to cool to room temperature before separating the material from glass.

3.3.1   Alternative Procedure

Since the glass transition temperature (Tg) of some polymers is below room temperature, where heating above Tg is not applicable, crimp can be induced by using a weak solvent to induce crystallinity change, such as ethanol. For PLLA and PCL (polycaprolactone), incubating in 95% ethanol at 37 °C for 2 days has been shown to induce crystallinity change and fiber crimping [6].

3.4   Material Validation

3.4.1   Fiber Morphology

To verify fiber morphology after electrospinning and treatment, scanning electron microscopy (SEM) is crucial (Fig. 4). However, in the case of PLLA nanofibers, to prevent fiber morphology changes induced by the energy generated during sputtering or high-intensity electric fields at high magnification, it is essential to

(A) As-spun   (B) Heat-treated

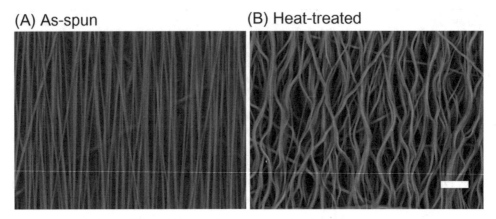

Fig. 4 PLLA fibers as spun (**a**) and after heat treatment (**b**). Scale bar equals 20 μm

use a low-field version that does not require surface coating, such as the Hitachi TM3000. Material thickness can also be assessed using the SEM.

Fiber quantification can include diameter and alignment, which can be assessed by image processing software, such as ImageJ [9]. A number of studies quantified fiber crimp by assessing the wavelength and amplitude; however, we find the measurement subjective in our system. To quantify fiber morphology change after heating, fiber path (L_f) and end-to-end fiber length (L_o) are measured using the NeuronJ plug-in in Image J [10]. Crimpness is defined as the difference in ratio of fiber path and end-to-end fiber length:

$$Crimpness = \frac{\left(L_f - L_o\right)}{L_o}$$

where higher values indicate wavier structures.

3.4.2 Mechanical Properties

Due to the crimp morphology of the fibers, the heat-treated material exhibits changes in its stress-strain response. We use a Bose Electroforce 5500 to apply uniaxial tensile loading at 0.02 mm/s and the stress is recorded with a 3 kg load cell until rupture, based on previously published parameters [8]. As shown in Fig. 5, heat-treated material has a larger toe region and extended transition strain.

3.4.3 Cell Seeding in the Electrospun Fibers

In addition to substrate morphology (dimensionality, topography, etc.), a great variety of additional factors influence cell-material interactions in general, such as chemical composition, mass transport for nutrients and waste, as well as cell-cell interactions [11–13]. The following protocol describes a basic seeding process that results in uniform cell seeding at subconfluent density, which

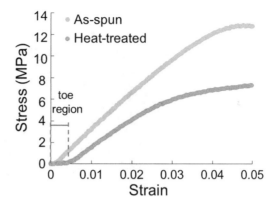

Fig. 5 Representative stress-strain response of as-spun and heat-treated materials

allows cell spreading and visualization of cell morphology while providing sufficient material for gene expression analysis. The following protocol is based on the material size of 1 × 2 cm processed in a 35 mm cell culture dish (or in a 6-well plate).

1. Sterilize materials in fresh hydrogen peroxide (enough to cover, approximately 1 mL) in a biological safety hood for 1 h, while UV is turned on. Rinse with PBS three times manually using a serological pipette. Using ethanol may induce fiber morphology changes. Conduct this and all the following steps in a biological safety hood with sterile reagents and consumables.

2. To facilitate protein binding, on the day of cell seeding, incubate the sample in 1 mL freshly sterile-filtered 2 mg/mL of dopamine solution in Tris buffer (pH 8.5) for 10 min at room temperature. Rinse with PBS three times.

3. Incubate the material in 100 µL of 25 µg/mL of fibronectin solution (in PBS) for 2 h at room temperature. Rinse with PBS three times.

4. On one side of the material, evenly distribute 45 µL of cell suspension, using a pipette, which contains 2 × 10^5 cells and place in a CO_2 incubator. The volume of cell suspension is kept at a minimum to prevent overflow. Add 15 µL of culture media (*see* **Note 4**) every 15 min for an hour to prevent drying.

5. Turn over the material and repeat **step 4** for the other side. At the end of the seeding process, flood the culture dish with 2.5 mL of media.

6. At the end of the culture period, cells can be visualized on confocal microscopy for morphology (for example, Fig. 6) or processed for gene expression or biochemical content and mechanical analysis (*see* **Note 5**).

As-spun

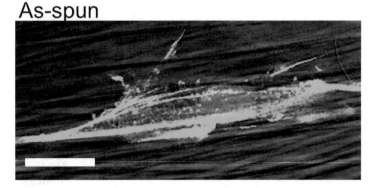

Heat-treated

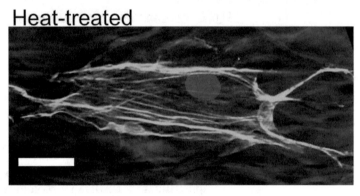

Fig. 6 Representative cell morphology in the as-spun and heat-treated materials. Scale bar equals 20 µm

4   Notes

1. Weighing the HFP in the syringe allows for more accurate measurements and prevents accumulation of PLLA at the tip. Since the PLLA-HFP solution is very viscous, making the solution in the syringe minimizes reagent loss from transferring.

2. PLLA concentration, electrical field strength, collector rotating speed, and flow rate can be optimized to achieve desired fiber morphology, alignment, and diameter.

3. Duration of the process depends on the size of the collector, the range of needle movement if using a fanner, and the desired material thickness. For the final material size of 12 × 25 cm and a thickness of 70 µm, electrospinning would take approximately 3 h.

4. Depending on the cell type, culture medium formulations may change. For basic mesenchymal stem cell (MSC) cultures, we used low-glucose DMEM supplemented with 10% FBS and antibiotics.

5. PLLA may interfere with RNA extraction using organic solvents. To improve RNA quality, a filter-based system is recommended (e.g., QuickRNA Mini Prep from Zymo Research).

Acknowledgments

This work was supported by the Taiwanese Ministry of Science and Technology (MOST-104-2221-E-002-107) and National Health Research Institute (NHRI-EX105-10411EI).

References

1. Lake SP, Miller KS, Elliott DM, Soslowsky LJ (2009) Effect of fiber distribution and realignment on the nonlinear and inhomogeneous mechanical properties of human supraspinatus tendon under longitudinal tensile loading. J Orthop Res 27(12):1596–1602

2. Cheng X, Gurkan UA, Dehen CJ, Tate MP, Hillhouse HW, Simpson GJ, Akkus O (2008) An electrochemical fabrication process for the assembly of anisotropically oriented collagen bundles. Biomaterials 29(22):3278–3288

3. Freeman JW, Woods MD, Laurencin CT (2007) Tissue engineering of the anterior cruciate ligament using a braid-twist scaffold design. J Biomech 40(9):2029–2036

4. Rwei SP, Lin YT, Su YY (2005) Study of self-crimp polyester fibers. Polym Eng Sci 45(6):838–845

5. Caves JM, Kumar VA, Xu W, Naik N, Allen MG, Chaikof EL (2010) Microcrimped collagen fiber-elastin composites. Adv Mater 22(18):2041–2044

6. Chao P-hG, Hsu H-Y, Tseng H-Y (2014) Electrospun microcrimped fibers with nonlinear mechanical properties enhance ligament fibroblast phenotype. Biofabrication 6(3):035008

7. Teo WE, Ramakrishna S (2006) A review on electrospinning design and nanofibre assemblies. Nanotechnology 17(14):R89

8. Baker BM, Gee AO, Metter RB, Nathan AS, Marklein RA, Burdick JA, Mauck RL (2008) The potential to improve cell infiltration in composite fiber-aligned electrospun scaffolds by the selective removal of sacrificial fibers. Biomaterials 29(15):2348–2358

9. Abramoff MD, Magelhaes PJ, Ram SJ (2004) Image processing with ImageJ. Biophoton Int 11(7):36–42

10. Meijering E, Jacob M, Sarria JCF, Steiner P, Hirling H, Unser M (2004) Design and validation of a tool for neurite tracing and analysis in fluorescence microscopy images. Cytometry A 58A(2):167–176

11. Chahine NO, Albro MB, Lima EG, Wei VI, Dubois CR, Hung CT, Ateshian GA (2009) Effect of dynamic loading on the transport of solutes into agarose hydrogels. Biophys J 97(4):968–975

12. Waggett AD, Benjamin M, Ralphs JR (2006) Connexin 32 and 43 gap junctions differentially modulate tenocyte response to cyclic mechanical load. Eur J Cell Biol 85(11):1145–1154

13. Breidenbach AP, Dyment NA, Lu Y, Rao M, Shearn JT, Rowe D, Kadler KE, Butler DL (2015) Fibrin gels exhibit improved biological, structural and mechanical properties compared to collagen gels in cell-based tendon tissue engineered constructs. Tissue Eng Part A 21(3–4):438–450

Chapter 13

In Vitro Model of Macrophage-Biomaterial Interactions

Claire E. Witherel, Pamela L. Graney, and Kara L. Spiller

Abstract

Tissue engineering and regenerative medicine, facilitated by biomaterial-based therapies, hold promise for the repair, replacement, or regeneration of damaged tissue. The success or failure of all implanted biomaterials, ranging from stainless steel total joint replacements to naturally or synthetically derived skin grafts, is predominantly mediated by macrophages, the primary cell of the innate immune system. In an effort to better assess safety and efficacy of novel biomaterials, evaluating and understanding macrophage-biomaterial interactions is a necessary first step. Here, we describe the culture of macrophages on 3D biomaterials, such as decellularized human cortical bone or commercially available wound matrices, and subsequent analysis using gene expression and protein secretion to help understand how biomaterial properties may influence macrophage phenotype in vitro.

Key words Macrophage phenotype, Human macrophage, Inflammation, Biomaterial

1 Introduction

In normal tissue repair following injury, including the implantation of biomaterials, the body follows a cascade of well-established and overlapping events including hemostasis, inflammation, proliferation, and remodeling [1]. During the early hemostasis and inflammatory phases, innate immune cells called macrophages are recruited to the site of injury to facilitate the clearance of debris, bacteria, and foreign material [1, 2] while also inducing fibroblasts to lay down extracellular matrix for tissue repair and regeneration [3, 4]. In the event that macrophages are unable to degrade or digest a foreign material, like an implanted biomaterial, they fuse into multinucleated foreign body giant cells (FBGCs) [5] and promote the formation of a dense, avascular fibrous capsule surrounding the implant [6]. The formation of FBGCs and fibrous encapsulation are considered hallmarks of the foreign body response [7], and are associated with implanted biomaterial failure due to chronic inflammation and/or inhibiting implant integration with the surrounding tissue [1, 8]. On the other hand, macrophages

Kanika Chawla (ed.), *Biomaterials for Tissue Engineering: Methods and Protocols*, Methods in Molecular Biology, vol. 1758, https://doi.org/10.1007/978-1-4939-7741-3_13, © Springer Science+Business Media, LLC, part of Springer Nature 2018

are critical for successful wound healing and biomaterial integration [9, 10]. Ultimately, the distinct effects of macrophages on the success or failure of biomaterial-mediated tissue regeneration processes appear to be related to their rapidly shifting activation states or phenotypes [2, 11]. Macrophages exist on a spectrum of distinct phenotypes that have unique roles in a variety of regenerative processes, including angiogenesis [3, 4, 12, 13] and wound healing [5, 9]. These diverse phenotypes include pro-inflammatory and pro-angiogenic M1 macrophages [6, 12–14]; anti-inflammatory M2a macrophages, associated with extracellular matrix (ECM) deposition and fibrosis [7, 12, 15]; and M2c macrophages, which are associated with tissue remodeling [11, 16, 17]. In normal wound healing and tissue regeneration, macrophages sequentially exhibit M1- and M2-like phenotypes over time [18]. Impairment in the M1-to-M2 transition, such as prolonged M1 macrophage phenotype, has been implicated in numerous diseases and pathologies including healing chronic wounds [19], myocardial infarction [20], and central nervous system regeneration [21]. Moreover, the microenvironment, including soluble signals from other cell types, the extracellular matrix, and other properties of the implanted biomaterials themselves, has a major effect on macrophage behavior [22, 23]. Therefore, there is a significant need to understand how macrophages are modulated by biomaterials in regenerative medicine strategies.

Subtle differences in macrophage phenotype observed using in vitro culture and gene expression analysis have been useful in describing macrophage behavior in vivo [24]. Unfortunately, there are limited tools and standardized methods to extensively characterize macrophage behavior in response to biomaterials outside of in vivo animal models, which may be poor models of the human immune response due to poor correlation between human and animal orthologs [25]. Here, we present one possible tool to explore human macrophage-biomaterial interactions in vitro, which may be helpful in exploring mechanisms of action of novel biomaterials and further the development of a validated predictive screen. Using human macrophages, which can be derived from peripheral blood monocytes or the human monocytic THP1 cell line, we culture cells onto various 3D biomaterials over time. Primary macrophages are the more physiologically relevant cell source [26], but suffer from some practical limitations, including donor-to-donor variability and limited life span in culture [27, 28]. While THP1 monocytes are a well-characterized proliferative cell line, which make them advantageous for large-scale projects [29], their disadvantage is the controversial evidence of their ability to mimic primary human macrophage behavior [27, 30]. Another experimental design consideration is whether to examine the effects of biomaterials on unactivated macrophages or macrophages of a specific phenotype, the latter of which may be more physiologically relevant for applications in inflammatory conditions such as chronic wounds, where macrophages are stalled in an M1, pro-inflammatory phenotype [31]. A variety of analysis techniques may be used downstream to characterize cell behavior in response to the biomaterial, including gene expression, protein secretion, cell viability assays, histology, etc. In this chapter, we describe in detail our protocols for cell culture, seeding cells onto biomaterials, processing samples for gene expression, and protein secretion for the evaluation of macrophage-biomaterial interactions. Any biomaterial can be used in this assay, ranging from synthetic total joint replacements to decellularized tissues. In particular, we focus on 3D biomaterial scaffolds used in tissue engineering and regenerative medicine. These techniques have applications in describing how different 3D biomaterials can affect macrophage behavior, which is critical for successful biomaterial integration and tissue regeneration.

2 Materials

1. Equipment and supplies.
2. Heated water bath.
3. Centrifuge and microcentrifuge.
4. 15 or 50 mL conical tubes.
5. 50, 250, or 500 mL sterile filters.
6. Ultralow-attachment vessel, e.g., T75 or T25 flasks, 6-, 12-, 24-, or 96-well plates (purchased from Corning; see **Note 1**).
7. Standard tissue culture vessel, e.g., T75 or T25 flasks, 6-, 12-, 24-, or 96-well plates.
8. Cell scrapers with 3.0 cm blade (if culturing cells in T flasks) or 11 mm blade (for cell culture in well plates).
9. Standard light microscope for quantifying viable cells.
10. Aspirating pipets (glass or disposable).
11. Bead beater and stainless steel beads.
12. 2 mL Eppendorf tubes.
13. RNA extraction kit.
14. RNA/cDNA quantification ability (i.e., Spectroscopy, NanoDrop, NanoQuant).
15. cDNA synthesis kit.
16. Quantitative real-time polymerase chain reaction (RT-PCR) machine.
17. Thermocycler.
18. Microcentrifuge tubes.
19. Quantitative RT-PCR 96-well plates.

Table 1
Example gene expression markers of macrophage phenotype [12, 17]

M1	M2a	M2c
CCR7	CCL18	CD163
IL1b	CCL22	MMP7
TNFa	MRC1	VCAN
VEGF	TIMP3	

Table 2
Example proteins for ELISA of macrophage phenotype [12, 17]

M1	M2a	M2c
IL1b	TIMP3	MMP9
TNFa	PDGF-BB	MMP7
VEGF		

20. RT-PCR primers of specific interest (Table 1).

21. ELISA kit for proteins of specific interest (Table 2).

2.1 Components for Culturing Primary Human Monocytes/ Macrophages

1. Complete RPMI (cRPMI) culture media: RPMI 1640 with phenol and L-glutamine, supplemented with 10% heat-inactivated human serum and 1% penicillin/streptomycin.

2. Recombinant human macrophage colony-stimulating factor (MCSF): Follow the supplier's instructions for reconstitution and storage. Avoid multiple freeze-thaw cycles of cytokines and polarizing stimuli.

2.2 Components for Culturing THP-1 Cells

1. Complete RPMI (cRPMI) culture media: RPMI 1640 with phenol and L-glutamine, supplemented with 10% heat-inactivated fetal bovine serum and 1% penicillin/streptomycin.

2. Phorbol 12-myristate 13-acetate (PMA) reconstituted in 100% ethanol or 20 mM dimethyl sulfoxide (DMSO) to yield a 100 μg/mL solution (*see* **Notes 2** and **3**): Reconstituted PMA is typically stored at 100 μg/mL in aliquots of 10–100 μL and frozen at −20°C ahead of time.

2.3 Components for Polarizing Macrophages

1. Lipopolysaccharide (LPS), recombinant human interferon-γ (IFN-γ), recombinant human interleukin-4 (IL-4), recombinant human interleukin-13 (IL-13), and recombinant human interleukin-10 (IL-10): Follow the supplier's instructions for reconstitution and storage. Avoid multiple freeze-thaw cycles of cytokines and polarizing stimuli.

2.4 Components ***for RNA Extraction***	1. 70% Ethanol.
	2. RNase displace.
	3. TRIzol.
	4. Chloroform, biotechnology grade.
	5. 100% Ethanol, molecular biology grade.
	6. DNase- and RNase-free distilled water.

3 Methods

3.1 Protocol
for Differentiating
Primary Human
Monocytes
into Macrophages

Primary human monocytes are differentiated into unactivated macrophages (M0 macrophages) by the addition of 20 ng/mL MCSF for 5 days in vitro, with a media replenishment on day 3. On day 5, unactivated macrophages are polarized into different phenotypes (M1, M2a, or M2c macrophages, as described in Subheading 1) by the addition of polarizing stimuli for an additional 24–48 h.

All cell culture should be conducted with sterile reagents in a biosafety cabinet.

3.1.1 Day 0

1. Pre-warm all cRPMI media components in 37 °C water bath and sterile filter.

2. When using fresh cells, wash cells with cRPMI one time before placing into culture.

3. When using frozen cells, thaw cells and wash cells with cRPMI one time before placing into culture.

4. Prepare a centrifuge tube with approximately 10 mL of cRPMI.

5. Warm frozen vial in water bath for approximately 30 s to 1 min, taking care to avoid submerging screw cap in water bath water, until liquid is visible in the tube.

6. Immediately transfer half-thawed cells by pipetting any liquid from the vial into the centrifuge tube containing 10 mL of cRPMI. Transfer a small volume of fresh media back into the vial to thaw and collect remaining cells. This technique will ensure rapid thawing of cells and dilution of the storage media, which often contains DMSO and is toxic to cells when thawed.

7. Culture primary monocytes at a concentration of 1.0×10^6 cells/mL with 20 ng/mL MCSF (*see* **Note 4**).

8. Incubate cells at 37 °C and 5% CO_2 for 3 days.

3.1.2 Day 3

1. Three days after initial culture, transfer cell suspension to a sterile conical tube (*see* **Notes 5** and **6**) and add 5 mL fresh, pre-warmed cRPMI media back to each T25 flask to prevent

death of any adherent macrophages left in the flask. Adherent cells are healthy macrophages and should not be scraped at this point in order to prevent any additional cell loss or death caused by processing and washing them.

2. Centrifuge cell suspension at $400 \times g$ for 7–10 min.

3. Aspirate supernatant and resuspend cell pellet in 5 mL fresh, pre-warmed cRPMI media.

4. Add 20 ng/mL recombinant human MCSF to the cell suspension and transfer to a T25 flask containing 5 mL media, for a total volume of 10 mL per flask. Incubate cells at 37 °C and 5% CO_2 for an additional 2 days.

3.1.3 Day 5

1. Collect cells by gently scraping (*see* **Note 7**) and transfer cell suspension to a sterile conical tube (*see* **Notes 5** and **6**).

2. Count cells and quantify cell viability using standard trypan blue exclusion assays [32] (*see* **Note 8**).

3. If unactivated (M0) macrophages are desired for seeding onto a biomaterial, this can be initiated on day 5, and proceed to Subheading 3.4 (*see* **Note 9**). However, if polarized (M1, M2a, or M2c, as described in Subheading 1) macrophages will be seeded onto biomaterials, cells must first be polarized; proceed to Subheading 3.3.

3.2 Protocol for Differentiating THP-1 Cells into Macrophages

1. Thaw a vial of THP1 cells.

2. Prepare a centrifuge tube with approximately 10 mL of cRPMI.

3. Warm frozen vial in water bath for approximately 30 s to 1 min, taking care to avoid submerging screw cap in water bath water, until liquid is visible in the tube.

4. Immediately transfer half-thawed cells by pipetting any liquid from the vial into the centrifuge tube containing 10 mL of cRPMI. Transfer a small volume of fresh media back into the vial to thaw and collect remaining cells. This technique will ensure rapid thawing of cells and dilution of the storage media, which often contains DMSO and is toxic to cells when thawed.

5. THP-1 monocytes are cultured at a concentration of 2.0×10^5 to 4.0×10^5 cells/mL in standard tissue culture vessels, and subcultured when the concentration reaches 8.0×10^5 cells/mL. THP-1 monocytes are centrifuged at $130 \times g$ for 7 min, in either 15 or 50 mL conical tubes. Do not exceed the concentration recommended by the manufacturer of the cell line (typically 1.0×10^6 cells/mL), as cells will not proliferate.

6. For differentiation into macrophages, culture 2.0×10^5 to 4.0×10^5 cells/mL in an ultralow-attachment vessel (*see* **Note 9**).

Table 3
Cytokines for polarizing macrophages

Phenotype	Polarizing factors		
M0	20 ng/mL MCSF[a]		
M1	20 ng/mL MCSF[a]	100 ng/mL IFNy	100 ng/mL LPS
M2a	20 ng/mL MCSF[a]	40 ng/mL IL-4	20 ng/mL IL-13
M2c	20 ng/mL MCSF[a]	40 ng/mL IL-10	

[a]MCSF is not needed if working with THP-1-derived macrophages

7. Add 2 μL PMA (100 μg/mL) per 1 mL cell suspension to yield a final PMA concentration of 320 nM (*see* **Notes 2** and **3**).

8. Incubate cells at 37°C and 5% CO_2 for 16 h; do not exceed 16 h, as cells tend to aggregate after this time.

9. Following incubation with PMA, transfer cell suspension to a sterile conical tube and add one-half volume of fresh cRPMI media back to vessel, to prevent death of adherent cells, as described in Subheading 3.1.2.

10. Centrifuge cell suspension at $130 \times g$ for 7 min.

11. Discard supernatant into hazard waste container, and resuspend cell pellet in fresh cRPMI media.

12. Repeat wash **steps 5–7** one to two additional times to fully remove all PMA from cells and vessel.

13. Transfer cell suspension to ultralow-attachment vessel and add polarizing factors, as needed, according to Table 3 (*see* **Notes 9** and **10**). If seeding biomaterials with polarized cells, incubate cells as described in Subheading 3.3, in the absence of MCSF (*see* **Note 11**), and then proceed to Subheading 3.4 (*see* **Notes 12**). Alternatively, if seeding unactivated THP1-derived macrophages (M0) on biomaterials, proceed directly to Subheading 3.4.

3.3 Protocol for Polarizing Primary, Unactivated Macrophages

1. Centrifuge unactivated (M0) macrophages at $400 \times g$ for 7–10 min.

2. Aspirate supernatant and resuspend cell pellet in pre-warmed cRPMI media.

3. Seed cells at 1.0×10^6 cells/mL in ultralow-attachment vessel. A minimum of one well or flask is required per macrophage phenotype, excluding technical replicates.

4. Polarize macrophages to the desired phenotype by addition of polarizing stimuli according to Table 3 (*see* **Note 9**).

5. Incubate cells in polarizing factors at 37°C and 5% CO_2 for 48 h (*see* **Notes 12**). In the event that polarized macrophage cell-only condition media is desired, *see* **Note 13**.

6. Detach macrophages by gentle scraping (*see* **Note 7**) and centrifuge the cell suspension at 400 × g for 7–10 min (*see* **Notes 5** and **6**).

7. Collect the supernatant/conditioned media, aliquot (*see* **Note 14**), and store at −80°C for future analysis or experiments. Protein secretion can be detected using commercially available enzyme-linked immunosorbent assay (ELISA) kits, according to the manufacturer's instructions, or conditioned media can be used to determine the effects of secreted proteins on other cell types.

3.4 Protocol for Seeding Macrophages on a Biomaterial (Fig. 1)

1. 3D biomaterials must be sterile before seeding them with cells (i.e., performed by soaking scaffolds in 70% ethanol for 30 min to 1 h and subsequently washing at least three times with sterile phosphate-buffered saline (PBS), gamma irradiation, ethylene oxide gas, etc.). Prepare biomaterials for cell seeding by soaking them in media at 37°C and 5% CO_2 for approximately 30 min to 1 h (*see* **Note 16**).

2. Resuspend macrophages in an appropriate volume of cRPMI media to seed an optimal concentration of cells for your application (*see* **Notes 17** and **18**).

3. Aspirate media from biomaterials and seed optimal volume of cell suspension directly onto the biomaterial. Cell-only controls are important in experimental design (*see* **Note 9**) and the methods for generating these are described in Subheading 3.3.

4. Incubate scaffolds at 37°C and 5% CO_2 for 30–60 min to allow cell attachment, but do not exceed 1 h of incubation. Prolonged incubation times can lead to cell death due to evaporation of the media from cell-seeded biomaterials.

5. Post-incubation, add a fixed volume of cRPMI media to completely submerge each sample in your chosen vessel. If working with primary macrophages, cRPMI media should be supplemented with 20 ng/mL MCSF. Some macrophages may not fully adhere to the biomaterial during the incubation time; if this is a concern (i.e., macrophages non-adherent to the biomaterial may influence macrophages seeded on the biomaterial via paracrine signaling and vice versa), *see* **Note 19**.

6. If using polarized macrophages, additional polarizing factors (Table 3) can be added to maintain macrophage phenotype, if relevant for the study (*see* **Note 20**).

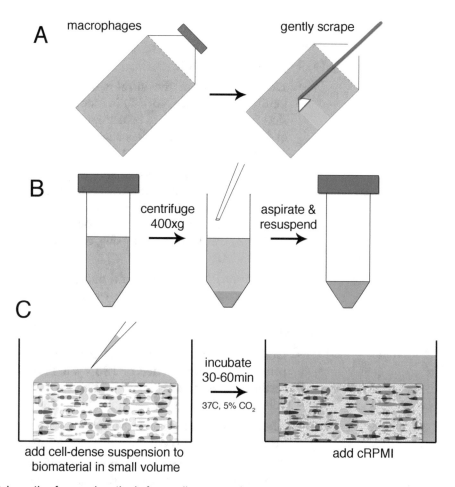

Fig. 1 Schematic of general methods for seeding macrophages onto biomaterials in vitro. (**a**) First, differentiated or polarized macrophages are gently scraped from the flask or well. (**b**) Macrophages are centrifuged and resuspended in a small volume, and (**c**) seeded in a small volume of media and allowed to infiltrate a 3D biomaterial for 30–60 min before adding culture medium. Constructs are cultured for desired periods of time from several days to weeks, with media changes routinely performed every 3–4 days. Conditioned media is collected for analysis or further experiments, and samples are collected and homogenized prior to RNA extraction

7. Incubate samples at 37°C and 5% CO_2, and replenish media every 3–4 days, as needed.

3.5 Protocol for Characterizing Macrophage Behavior Using RT-PCR and ELISA

In an effort to understand how biomaterials may affect macrophage behavior in vitro, gene expression and protein secretion of several identified and validated markers are employed (Tables 1 and 2). By analyzing these data, it is possible to determine if a biomaterial may promote a more inflammatory or anti-inflammatory macrophage phenotype, which is important for tissue regeneration.

1. Prepare Samples for ELISA and RNA Extraction for Later Gene Expression Analysis:

(a) For cell-only controls:

- Detach cells from the ultralow-attachment surface by gentle scraping (*see* **Note 7**) in media, pipet cell suspension into an appropriate centrifuge tube, and centrifuge at $400 \times g$ for primary cells or $130 \times g$ for THP1 cells for approximately 7 min.

- Collect conditioned media and aliquot into separate centrifuge tubes to minimize subsequent freeze-thaw cycles (*see* **Note 14**). Store conditioned media at −80°C for future analysis. Protein secretion of selected markers (Table 2) can be detected using commercially available ELISA kits.

- For future RNA isolation and purification, lyse the cell pellet in appropriate volume of lysis buffer (*see* **Note 15**) and store lysate at −80°C.

(b) For cell-seeded biomaterials:

- Collect conditioned media and centrifuge at $400 \times g$ if primary cells were seeded or $130 \times g$ if THP1 cells were seeded for approximately 7 min, to exclude any nonadherent cells (*see* **Note 21**).

- Aliquot conditioned media into new tubes (*see* **Note 14**). Store conditioned media at −80°C for future analysis.

- Transfer each sample into a microcentrifuge tube containing 1 mL of ice-cold TRIzol in a chemical fume hood (*see* **Note 22**). Samples can be stored in TRIzol at −80°C for future RNA extraction.

(c) Prior to RNA extraction, homogenization may be required and must be performed with care to ensure high-quality yield (*see* **Note 23**).

- For cells: Cells can be lysed via vigorous micropipetting.

- Tissues or cell-containing biomaterials must be homogenized to fully and thoroughly break up the material and cells for effective RNA extraction. This can be performed with a probe homogenizer or bead beater. Probe homogenizers must be adequately cleaned between each sample to ensure no RNA or DNA contamination (*see* **Notes 23a** and **23b**).

  – If samples are frozen, briefly vortex or invert tubes and incubate samples at room temperature for 1–5 min to ensure that they are adequately thawed.

  – Add 0.2 mL chloroform per 1 mL of TRIzol Reagent used. Cover the sample tightly, shake vigorously by hand for 15 s (do not vortex), and incubate for 2–3 min at room temperature.

- Centrifuge the resulting mixture at $12,000 \times g$ for 15 min at 4°C. A clear phase separation should be visible, with the aqueous (clear) phase forming the top layer, and a white/pale interphase and the organic (pink) phase forming the bottom layer.

- Transfer the maximum amount of the top, clear, aqueous phase, without disrupting the interphase, to a fresh tube (*see* **Note 24**). Indicate the volume collected on the tube for the next step. Proceed with RNA purification using commercially available kits according to the manufacturer's instructions (*see* **Note 25**).

(d) Quantify the RNA using spectroscopy and confirm the purity by measuring the absorbance at 230, 260, and 280 nm (*see* **Note 26**).

(e) For small quantities of RNA, *see* **Note 27**. All purified RNA should be treated for possible DNA contamination (confirmed or unconfirmed via RNA quantification method). Lastly, purified RNA is then reverse transcribed to create complementary DNA (cDNA) for use in quantitative RT-PCR (*see* **Note 28**). For specific recommendations on performing RT-PCR and subsequent analysis, *see* **Note 29**.

4   Notes

1. Ultralow-attachment surfaces are recommended when culturing primary human macrophages and PMA-activated THP-1 cells to minimize activation of the cells, which can result from culture and attachment on standard tissue culture plastic [33].

2. PMA is a carcinogen. It is recommended to layer two pairs of nitrile gloves when handling PMA and use extreme caution. All liquids containing PMA, and any material that comes in contact with PMA, should be discarded in appropriate hazardous waste containers. All vessels containing PMA should be labeled accordingly.

3. We recommend reconstituting PMA in 100% ethanol, molecular biology grade.

4. We routinely add 8.0×10^6 to 10.0×10^6 primary monocytes in 10 mL of media to each T25 flask, which can be scaled for a T75 flask if working with large quantity of cells.

5. If multiple flasks are used for macrophage culture, the cell suspensions may be pooled for centrifugation.

6. Centrifugation in 15 mL conical tubes is optimal for pelleting primary macrophages, as we have experienced significant cell loss when aspirating the supernatant from cells pelleted in 50 mL conical tubes.

7. It is critical to visualize by light microscopy that the cell monolayer has been detached to maximize cell yield. Macrophages tend to migrate to the perimeter and edges of the flasks or wells, so extra care should be taken to detach cells located in these areas.

8. We routinely achieve a 40–60% yield of macrophages on day 5, relative to the number of monocytes cultured on day 0.

9. Inclusion of M0, M1, M2a, and M2c cell controls, i.e., those not seeded on biomaterials, provides a reference for determining the response of macrophages to the biomaterial. For example, the gene expression or protein secretion profiles of macrophages cultured on a biomaterial can be directly compared to the gene expression of each macrophage phenotype cultured under standardized conditions in vitro. Or, if a specific phenotype is seeded onto biomaterials, it is imperative that a cell-only control of the seeded phenotype is generated for data analysis.

10. We have also successfully differentiated THP-1 cells at concentrations of 0.5×10^6 to 1.0×10^6 cells/mL when working with larger cell populations. It is essential to use ultralow-attachment flasks during this step to minimize cell attachment. When ultralow-attachment flasks are not used, aggressive scraping is required which can lead to inadvertent cell activation or damage.

11. MCSF is not needed for differentiating THP-1 cells into unactivated macrophages, which is achieved using PMA.

12. Macrophages can be polarized for 24–72 h. We routinely polarize macrophages for 48 h.

13. If conditioned media of cell-only controls is desired following polarization, resuspend cells at a concentration of 1.0×10^6 cells/mL in cRPMI media without additional polarizing factors and culture cells for an additional 24 h to ensure that only endogenous cell cytokine production is collected.

14. Conditioned media should be aliquotted in 100–500 μL volumes to avoid multiple freeze-thaw cycles and provide easy use for multiple assays. The volume required for protein secretion analysis will vary based on the sensitivity of the assay.

15. We routinely seed cells on biomaterials (i.e., collagen sponges, demineralized human cortical bone) that have been hydrated with sterile media to facilitate more uniform cell seeding. However, cells can also be seeded on sterile, dry biomaterials,

which may facilitate cell attachment especially for very hydrophilic materials. Extra care will be necessary to ensure that cells do not dry out if seeding onto sterile, dry scaffolds.

16. It is necessary to predetermine the optimal cell seeding volume based on the biomaterial of interest. Avoid volumes that lead to spillover from the biomaterial into the well. We have routinely seeded 1.0×10^6 cells in 10–20 μL onto 5–10 mm × 1–2 mm biopsy punches of spongelike biomaterials.

17. A minimum of 0.5×10^6 cells per sample typically provides enough RNA for gene expression characterization.

18. It is possible for cells to float off the biomaterial during incubation time or after the addition of cRMPI media to the well. If this is a concern, use sterile forceps to transfer the cell-seeded biomaterials to new wells to ensure that only adherent cells are included in the sample.

19. During experimental design it is important to consider that the addition of cytokines may affect any other cell types (i.e., cell-containing biomaterials) included in the assay.

20. We recommend that cells be lysed in the lysis buffer provided in kits, as using the standard TRIzol-chloroform method may yield low-quality RNA for cell-only samples. RNA from cell-seeded biomaterials should be extracted using TRIzol-chloroform followed by RNA purification as described above.

21. Be sure to carefully remove media from biomaterials seeded with macrophages, as small movements may disturb the sample and loosely adherent cells on the surface of the material.

22. TRIzol Reagent is a hazardous substance. Handle with caution. RNA extraction with TRIzol and subsequent RNA purification must be completed in a chemical fume hood.

23. We recommend that all cell-seeded biomaterials be homogenized in some fashion to ensure high-quality yield during RNA extraction. We have used a variety of methods, including a probe homogenizer or a bead beater. During homogenization, always keep samples on ice. It is important to ensure that the sample is thoroughly broken up into very small particles to ensure that any cells within the material are freed and lysed. It is appropriate to move forward with RNA extraction even if the sample is not fully liquefied and there are remaining biomaterial particles.

(a) For the probe homogenizer, samples are homogenized between 15,000 and 20,000 rpms for several (at least three) short rounds (no greater than 15 s each), to ensure that the sample does not heat up and degrade the RNA. The homogenizer bit must be cleaned with RNase Displace, 70% ethanol, and RNase/DNase-free water three times in between each sample.

(b) For the bead beater, samples are placed into 2 mL Eppendorf tubes along with three to four 2.3 mm stainless steel beads and homogenized for several (at least three) short rounds (no greater than 15 s) to ensure that the sample does not heat up and degrade the RNA. Fresh beads are used for each sample to ensure no cross contamination of RNA or DNA between samples. Beads are kept in the tube during centrifugation with chloroform.

24. To avoid collecting the interphase, we typically collect the aqueous phase in 100 μL increments for the first few samples to estimate the maximum amount of volume that can be collected without contamination. Once this volume is established, it is recommended that a larger one-time volume be removed for efficiency.

25. We have found that the QIAGEN RNeasy Mini Kit works well when purifying RNA isolated from TRIzol.

26. When determining the concentration of RNA in a sample, the absorbance measurements taken at 230, 260, and 280 nm are important for indicating the purity of the sample. A ratio of 260/280 from 1.8 to 2.1 is considered pure, with numbers less than 1.8 indicating possible DNA contamination and numbers greater than 2.1 indicating possible protein or phenol/chloroform contamination. A 260/230 ratio between 2.0 and 2.2 is considered pure, with lower ratios indicating contaminants. Generally, the use of these ratios to determine which samples to include in your analysis depends on the downstream analysis planned. For example, if you plan analysis that may be sensitive to DNA contamination, like RT-PCR, it is very important to ensure that your samples are considered pure.

27. For small quantities of RNA (less than 1 μg), when processing for DNA contamination and reverse transcription (*see* **Note 29**), the manufacturer's protocols may be doubled or tripled to ensure that maximum RNA is processed downstream.

28. We typically use a DNase kit to purify the RNA (DNase I, Amplification Grade, Life Technologies), then reverse transcribe the RNA to create the template for RT-PCR analysis (High Capacity cDNA Reverse Transcription Kit, Life Technologies), and perform quantitative RT-PCR using Fast SYBR Green (Life Technologies), all according to the manufacturers' protocols.

29. Qiagen and Thermo Scientific have particularly helpful online manuals and background information on RT-PCR basics that can be found here: https://www.qiagen.com/us/resources/molecular-biology-methods/pcr/ and https://www.thermo-fisher.com/us/en/home/life-science/pcr/real-time-pcr/

qpcr-education.html. Several articles have been published on other important considerations when performing RT-PCR including the importance of reference gene selection [34] and analysis techniques [35].

References

1. Diegelmann RF (2004) Wound healing: an overview of acute, fibrotic and delayed healing. Front Biosci 9:283

2. Leibovich SJ, Ross R (1975) The role of the macrophage in wound repair. Am J Pathol 78:71–100

3. Leibovich SJ, Ross R (1976) A macrophage-dependent factor that stimulates the proliferation of fibroblasts in vitro. Am J Pathol 84:501–514

4. Khalil N, Bereznay O, Sporn M et al (1989) Macrophage production of transforming growth factor-beta and fibroblast collagen synthesis in chronic pulmonary inflammation. J Exp Med 170:727–737

5. McNally AK, Anderson JM (1995) Interleukin-4 induces foreign body giant cells from human monocytes/macrophages. Am J Pathol 147:1487–1499

6. Clark AE, Hench LL, Paschall HA (1976) The influence of surface chemistry on implant Interface histology: a theoretical basis for implant materials selection. J Biomed Mater Res 10:161–174

7. Anderson J (2001) Biological response to materials. Annu Rev Mater Res 31:81–110

8. Anderson JM, Rodriguez A, Chang DT (2008) Foreign body reaction to biomaterials. Semin Immunol 20:86–100

9. Mirza R, DiPietro LA, Koh TJ (2009) Selective and specific macrophage ablation is detrimental to wound healing in mice. Am J Pathol 175:2454–2462

10. Roh JD, Sawh-Martinez R, Brennan MP et al (2010) Tissue-engineered vascular grafts transform into mature blood vessels via an inflammation-mediated process of vascular remodeling. Proc Natl Acad Sci 107:4669–4674

11. Yancopoulos GD, Davis S, Gale NW et al (2015) Vascular-specific growth factors and blood vessel formation. Nature 407:242–248

12. Spiller KL, Anfang RR, Spiller KJ et al (2014) The role of macrophage phenotype in vascularization of tissue engineering scaffolds. Biomaterials 35:4477–4488

13. Willenborg S, Lucas T, van Loo G et al (2012) CCR2 recruits an inflammatory macrophage subpopulation critical for angiogenesis in tissue repair. Blood 120:613–625

14. Mirza RE, Fang MM, Ennis WJ et al (2013) Blocking interleukin-1b induces a healing-associated wound macrophage phenotype and improves healing in type 2 diabetes. Diabetes 62:2579–2587

15. Xue J, Sharma V, Hsieh MH et al (2015) Alternatively activated macrophages promote pancreatic fibrosis in chronic pancreatitis. Nat Commun 6:1–11

16. Lolmede K, Campana L, Vezzoli M et al (2009) Inflammatory and alternatively activated human macrophages attract vessel-associated stem cells, relying on separate HMGB1- and MMP-9-dependent pathways. J Leukoc Biol 85:779–787

17. Spiller KL, Nassiri S, Raman P et al (2015) Discovery of a novel M2c macrophage gene expression signature indicates a major role in human wound healing. Wound Repair Regen 23(2):A40

18. Arnold L, Henry A, Poron F et al (2007) Inflammatory monocytes recruited after skeletal muscle injury switch into antiinflammatory macrophages to support myogenesis. J Exp Med 204:1057–1069

19. Nassiri S, Zakeri I, Weingarten MS et al (2015) Accepted article preview: published ahead of advance online publication. J Investig Dermatol 135:1700–1703

20. Zhou L-S, Zhao G-L, Liu Q et al (2015) Silencing collapsin response mediator protein-2 reprograms macrophage phenotype and improves infarct healing in experimental myocardial infarction model. J Inflamm 12:1–12

21. Miron VE, Boyd A, Zhao J-W et al (2013) M2 microglia and macrophages drive oligodendrocyte differentiation during CNS remyelination. Nat Neurosci 16:1211–1218

22. Lavin Y, Winter D, Blecher-Gonen R et al (2014) Tissue-resident macrophage enhancer landscapes are Shapedby the local microenvironment. Cell 159:1312–1326

23. Brown BN, Londono R, Tottey S et al (2012) Macrophage phenotype as a predictor of constructive remodeling following the implantation of biologically derived surgical mesh materials. Acta Biomater 8:978–987

24. Xue J, Schmidt SV, Sander J et al (2014) Transcriptome-based network analysis reveals a Spectrum model of human macrophage activation. Immunity 40:274–288

25. Seok J, Warren HS, Cuenca AG et al (2013) Genomic responses in mouse models poorly mimic human inflammatory diseases. Proc Natl Acad Sci 110:3507–3512

26. Heil TL, Volkmann KR, Wataha JC et al (2002) Human peripheral blood monocytes versus THP-1 monocytes for in vitro biocompatibility testing of dental material components. J Oral Rehabil 29:401–407

27. Spiller KL, Wrona EA, Romero-Torres S et al (2016) Differential gene expression in human, murine, and cell line-derived macrophages upon polarization. Exp Cell Res 347(1):1–13

28. Aldo PB, Craveiro V, Guller S et al (2013) Effect of culture conditions on the phenotype of THP-1 monocyte cell line. Am J Reprod Immunol 70:80–86

29. Qin Z (2012) The use of THP-1 cells as a model for mimicking the function and regulation of monocytes and macrophages in the vasculature. Atherosclerosis 221:2–11

30. Zheng L, Martins-Green M (2007) Molecular mechanisms of thrombin-induced interleukin-8 (IL-8/CXCL8) expression in THP-1-derived and primary human macrophages. J Leukoc Biol 82:619–629

31. Sindrilaru A, Peters T, Wieschalka S et al (2011) An unrestrained proinflammatory M1 macrophage population induced by iron impairs wound healing in humans and mice. J Clin Investig 121:985–997

32. Strober W (2001) Trypan blue exclusion test of cell viability. In: The isolation and characterization of murine macrophages. Wiley, New York

33. Kelley JL, Rozek MM, Suenram CA et al (1987) Activation of human blood monocytes by adherence to tissue culture plastic surfaces. Exp Mol Pathol 46:266–278

34. Dheda K, Huggett JF, Chang JS et al (2005) The implications of using an inappropriate reference gene for real-time reverse transcription PCR data normalization. Anal Biochem 344:141–143

35. Livak KJ, Schmittgen TD (2001) Analysis of relative gene expression data using real-time quantitative PCR and the $2-\Delta\Delta CT$ method. Methods 25:402–408

Chapter 14

Synthesis of Self-Assembling Peptide-Based Hydrogels for Regenerative Medicine Using Solid-Phase Peptide Synthesis

E. Thomas Pashuck

Abstract

Peptide self-assembly is an important field in biomaterials in which short peptides are designed to aggregate into nanostructures that often form hydrogels. These peptides are typically made using solid-phase peptide synthesis (SPPS), a technique in which amino acids are added sequentially to a growing chain. This technique has been used to synthesize peptides with more than 100 amino acids. However, self-assembling peptides are designed to aggregate in solution, which often reduces coupling efficiency during synthesis and makes purification more difficult. Here, an outline of solid-phase peptide synthesis is provided, along with steps that can be used to improve the synthetic yield and purification of self-assembling peptides for regenerative medicine applications.

Key words Peptide, Solid-phase peptide synthesis, Self-assembly, Biomaterials, High-performance liquid chromatography

1   Introduction

Self-assembling peptides are a class of biomaterial that has seen significant interest in the field of regenerative medicine due to their injectability, biocompatibility, and ease of bioactive modification [1]. These peptides typically form β-sheets and self-assemble into high-aspect-ratio nanostructures that are capable of forming hydrogels under physiological conditions without chemical cross-linking [2]. The most widely studied peptides are generally less than 25 amino acids and are almost exclusively synthesized via solid-phase peptide synthesis (SPPS). Peptides of this length can usually be synthesized at high purity and without significant optimization and are easier to handle post-synthesis. In this technique, the growing peptide is attached to a solid support and amino acids are added sequentially through the iterative process of amino acid coupling and then the removal of the protecting group, as depicted

Kanika Chawla (ed.), *Biomaterials for Tissue Engineering: Methods and Protocols*, Methods in Molecular Biology, vol. 1758, https://doi.org/10.1007/978-1-4939-7741-3_14, © Springer Science+Business Media, LLC, part of Springer Nature 2018

1 Fmoc protected amino acid   **2** Activated amino acid ester

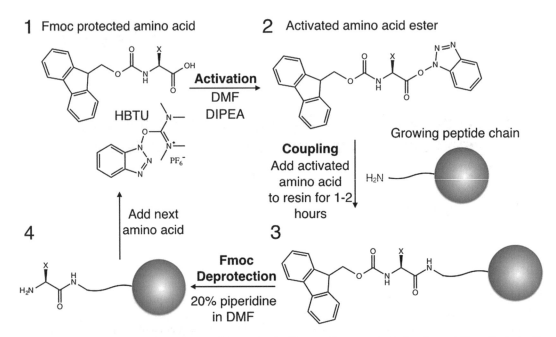

Fig. 1 The steps of Fmoc solid-phase peptide synthesis. First, an Fmoc-protected amino acid is activated with the coupling reagent, dissolved in dimethylformamide (DMF) and diisopropylethylamine to form the activated ester of the amino acid (Step 2). This is then added to the solid-phase resin with N-terminal amines present and placed on a wrist action shaker for 1–2 h (Step 3). Upon successful completion of the coupling the Fmoc-protecting group is removed with a 20% piperidine/80% DMF mixture, liberating the N-terminal amine on the growing peptide chain (Step 4), whereupon the cycle is repeated

in Fig. 1. In solid-phase peptide synthesis an amino acid with an Fmoc-protecting group on the amine is mixed with coupling reagents, such as N,N,N',N'-tetramethyl-O-($1H$–benzotriazol-1-yl)uronium hexafluorophosphate (HBTU) (Fig. 1, Step 1), to form a reactive activated ester (Fig. 1, Step 2). The solution is then added to the solid-phase support, which has free amines present on the end of the peptide chain, which react with the activated ester to form a new amide (or peptide) bond (Fig. 1, Step 3). Upon completion of the reaction, the Fmoc group protecting the amine on the added amino acid is removed (Fig. 1, Step 4) and the cycle is repeated. Solid-phase peptide synthesis has many benefits in that it is relatively inexpensive, can easily incorporate a variety of non-natural amino acids or functional groups, can be adjusted to the scale of most laboratory needs, and only requires basic knowledge of chemistry [3]. Self-assembling peptides, which are designed to aggregate in solution, can be challenging to synthesize through SPPS, as aggregation is the primary reason for reduced coupling efficiency during peptide synthesis. The propensity for assembling into larger structures can also be problematic for purification by high-performance liquid chromatography (HPLC). Fortunately, the field of peptide synthesis has developed a variety of ways to either prevent or reduce aggregation, which allows for the vast majority of difficult peptide sequences to be synthesized [4]. Utilizing these steps, the methodology for synthesizing a model peptide (palmitoyl-VVVAAAEEE molecule) is provided below [5].

2   Materials

2.1   Equipment and Glassware Required

1. Glass peptide synthesis vessel (Chemglass CG-1860) with screw-top cap with a PTFE seal, a coarse porous fritted glass resin support at the bottom with a stopcock and glass stem at the bottom.

2. Wrist-action shaker (Burrell Scientific).

3. Rotary evaporator (Buchi).

4. Büchner funnel with glass frit.

5. Side-arm flask (also called Büchner flask).

6. Rubber stopper placed in the top of the flask with a hole for the glass stem of the peptide synthesis vessel.

7. Vacuum pump/vacuum line.

8. Hot plate.

9. 50 mL Glass beaker.

10. 50 mL Conical tubes.

11. Sand.

12. Thermometer.

13. Disposable glass culture tube (8 or 10 mm in diameter).

14. Glass Pasteur pipettes.

15. Plastic wash bottles for solvent.

2.2   Peptide Synthesis and Cleavage

1. Solid support resin (see **Note 1**).

2. Fmoc-protected amino acids.

3. N,N,N',N'-Tetramethyl-O-($1H$–benzotriazol-1-yl)uronium hexafluorophosphate (HBTU) peptide coupling reagent (see **Note 2**).

4. N,N-diisopropylethylamine (DIPEA).

5. N,N-dimethylformamide (DMF) (see **Note 3**).

6. Dichloromethane (DCM).

7. Piperidine.

8. Kaiser test kit (see **Note 4**):

Kaiser test kits can either be purchased as premade kits or made using ninhydrin, ethanol, liquefied phenol, pyridine, and potassium cyanide.

9. Acetic anhydride.

10. Trifluoroacetic acid (TFA).

11. Triisopropylsilane (TIPS).

12. Dithiothreitol (DTT).

13. Diethyl ether.

2.3 Peptide Purification

1. Ultrapure water.

2. Acetonitrile.

3. Trifluoroacetic acid.

4. Ammonium hydroxide.

5. Ammonium formate.

3 Methods

3.1 Synthesis

1. Determine the scale of peptide synthesis and weigh out the appropriate amount of the peptide synthesis resin (*see* **Note 5**). Place the resin into the peptide synthesis vessel and add dichloromethane (DCM) for 10 min to swell the resin before synthesis. For our model peptide we will synthesize 0.5 mmol of peptide using a 0.5 mmol/g Rink-amide resin, for which 1 g of resin should be added to the synthesis vessel. Use the plastic wash bottle to add enough to fill the vessel to roughly one-third full and such that there is liquid DCM remaining in the vessel after the resin swells.

2. Place synthesis vessel onto the wrist-action shaker and shake for 10 min.

3. Fill a small glass beaker with sand place onto a hot plate and adjust the temperature such that the temperature in the sand is between 100 and 110 °C.

4. Attach a large Büchner flask to a vacuum line and place a stopper in the top with a hole that fits the glass stem of the peptide synthesis vessel. The solvent should always be emptied from the flask before it reaches the level of the side arm.

5. An "Fmoc Peptide Synthesis Schedule" (Fig. 2) is included which can be printed and filled out with the peptide sequence to easily track progress during peptide synthesis while listing the necessary steps. It is important to note that peptides are synthesized from the C-terminus to the N-terminus (C to N) while peptides are written from N to C, so for a given peptide sequence synthesis occurs from right to left as written. For our palmitoyl-VVVAAAEEE, the first coupling will be the glutamic acids (E) and it ends with the N-terminal palmitoyl functionalization.

Date started: _____   Date finished: _____   Page: _____ of _____

Fmoc Peptide Synthesis Schedule

Peptide Name/Number: _____

Resin: _____   Scale: _____

Sequence (C-N)									
Date									
20% Pip/DMF(5-10 min)									
20% Pip/DMF(5-10 min)									
DMF (2x)									
DCM (2x)									
Kaiser Test									
Coupling (1-2 hours)									
DMF (2x)									
DCM (2x)									
Kaiser test									
If needed									
Capping (5 min)									
Capping (5 min)									
DCM (4x)									
DMF (2x)									

Coupling solution: 4:3.95:6 (molar equivalents) amino acid:HBTU:DIPEA in DMF
Capping solution: 10:2.5:100 (v/v/v) acetic anhydride:DIPEA:DMF

Fig. 2 Fmoc peptide synthesis schedule. The peptide sequence is filled from the C-N direction (note that protein/peptide sequences are written N-C; thus the RGDS peptide would be filled in SDGR from left to right, as it is synthesized) and each step is checked off as it is completed, going from top to bottom and then left to right

6. Take the synthesis vessel off the shaker and use the side-arm flask to remove the dichloromethane with the vacuum. Add 5–10 mL of 20% piperidine in DMF (pip/DMF) per gram of resin to the synthesis vessel to cleave the Fmoc group from the resin.

7. Shake vessel for 5 min, use the vacuum to remove the pip/DMF, add fresh pip/DMF solution, and place on shaker again for 5 min. Use the vacuum to remove the pip/DMF and wash the resin by filling the synthesis vessel one-third full with DMF, then removing it with the vacuum. Repeat this with DMF and then wash 2× with DCM.

8. Add one drop of each of the three Kaiser test solutions into a glass scintillation vial and use glass Pasteur pipette (or spatula) to remove a small amount of the resin from the peptide synthesis vessel and into the glass vial. Place this vial into the heated sand bath and wait for approximately 60 s. In the presence of a free amine after Fmoc deprotection, the Kaiser test solution should turn a dark purple color as seen in Fig. 3. If the Kaiser

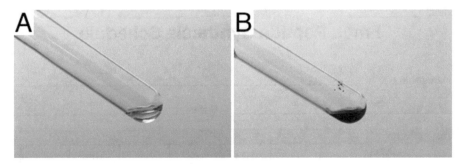

Fig. 3 Examples of a negative (**a**) and positive (**b**) Kaiser test

test does not turn dark purple after Fmoc deprotection, *see* **Note 6**.

9. Upon successful deprotection, add DCM to swell the resin for 3–5 min.

10. Weigh out 4 equivalents of the amino acids and 3.95 equivalents of HBTU and place into 50 mL centrifuge tube or glass vial (*see* **Note 7**). For glutamic acid at the 0.5 mmol scale, add 851 mg of Fmoc-Glu(OtBu) and 749 mg of HBTU in a 50 mL conical tube.

11. Fill the tube with either DMF or NMP (*see* **Note 3**), adding enough to completely dissolve the HBTU and amino acid. Typically, 10–15 mL of solvent is added per gram of resin used, enough to coat all areas of the synthesis vessel during wrist-action shaking. This is followed by the addition of 6 equivalents of DIPEA to the vial, which is 520 μL for a 0.5 mmol scale synthesis.

12. After a minute, the HBTU will have activated the carboxylic acids on the amino acid and the solution can be added to the resin. In order to prevent areas of incomplete coupling, add enough solvent to ensure that all parts inside the vessel come into contact with the solution during shaking. For most peptide couplings a significant majority of the amines have been reacted within the first few minutes, but shaking for 1–2 h is more common to ensure that the reaction goes to completion.

13. After 1–2 h, remove the solution inside the synthesis vessel into the Büchner funnel and wash the resin twice with DMF and twice with DCM.

14. Perform a Kaiser test to check for the presence of free amines. If a negative Kaiser test is seen, as depicted in Fig. 3, this coupling step has been completed successfully. Go to **step 6** and repeat **steps 6–14** with the next amino acid. For the palmitoyl-VVVAAAEEE peptide, two more glutamic acids are added sequentially, followed by three alanines and three valines. If the Kaiser test is not completely negative, *see* **Note 8**.

3.2 Cleavage

Once peptide synthesis has been completed, it is typically placed into a cleavage solution which has the dual role of freeing the peptide from the solid-phase support and removing any protecting groups on the side chains of the peptide. The ideal cleavage solution removes all the desired protecting groups while preventing unwanted side reactions. The most commonly used cleavage solution is 95% trifluoroacetic acid (TFA), with 2.5% H_2O and 2.5% triisopropylsilane (TIPS). However, peptides containing cysteine, methionine, and tryptophan will require an extra additive (*see* **Note 11**). It is important to note that TFA is corrosive and should be handled with caution. Furthermore, check the cap of the peptide synthesis vessel to ensure that the PTFE seal in the cap is intact without any holes. If the PTFE is missing or contains holes the TFA will dissolve the cap.

After the final Fmoc deprotection or capping steps:

1. Wash the resin 2× with DMF and 3× with DCM.

2. At this point the N-terminal amine can be modified or capped (acetylated) before peptide cleavage. A palmitoyl modification, as is found on the palmitoyl-VVVAAAEEE peptide, is added in a similar manner to a standard amino acid. For a 0.5 mmol synthesis, 512 mg of palmitic acid is dissolved in 15–20 mL of DMF with 749 mg of HBTU and 520 µL of DIPEA. For hydrophobic modifications, such as palmitoylation, a small amount of DCM may be needed to fully solubilize the palmitic acid. This is then reacted for 1–2 h on the wrist-action shaker and a Kaiser test is done to check that the coupling went to completion. The resin is then washed several times with DMF and DCM before moving onto the next step. Instruction for capping the N-terminus can be found in **Note 8**.

3. After the final DCM washing step, add approximately 10 mL of cleavage solution per gram of resin.

4. Place on shaker for 2 h. If arginine is present in the peptide sequence shake for 3 h to fully remove the 2,2,4,6,7-pentamethyldihydrobenzofuran-5-sulfonyl (Pbf)-protecting group on arginine.

5. After the cleavage is complete, empty the contents of the shaker vessel into a round-bottomed flask and rinse the resin 3× with DCM to remove any peptide left in the vessel.

6. Place the round-bottomed flask with free peptides in a TFA/DCM solution and use rotary evaporation for 5–10 min to remove as much of the solvent as possible.

7. Precipitate the peptide in cold diethyl ether. Add approximately 50 mL of diethyl ether per mmol of peptide.

8. The diethyl ether can be removed from the peptide by either filtering the peptide with a fritted Buchner funnel or using a

centrifuge to pellet the peptide in the bottom of a 50 mL conical tube.

9. Remove any remaining ether by leaving the tube uncapped overnight in a fume hood or under vacuum, before the peptide is solubilized for purification.

3.3 Purification

Most peptides are purified through reverse-phase high-performance liquid chromatography (HPLC). The peptide is injected onto a column and a gradient is run from mostly water to increasing concentrations of acetonitrile or methanol. A typical column contains a silica phase functionalized with hydrocarbon chains which sequesters the injected peptide until the water/acetonitrile concentration solubilizes the particular peptide and it leaves the column and flows past a detector into the fraction collector. Thus in HPLC, materials are purified based on differences in their solubility in water, as more hydrophobic peptides typically come out of the column ("eluted") at later times during a run.

Most preparative HPLC is done under slightly acidic conditions, where 0.1% TFA is added to water and acetonitrile to improve solubility of the peptides. However, for some peptides which contain few or no positively charged amino acids (lysine, arginine, and histidine), but do contain negatively charged ones (glutamic and aspartic acid), it may be easier to do purification under basic conditions. In this case 0.1% NH_4OH is added to the water and acetonitrile. It is important to ensure that the HPLC column can withstand high pH, as most standard silica columns will start to degrade under basic conditions.

General methodology for preparative HPLC

1. Set up an HPLC method with a gradient of 95% H_2O/5% acetonitrile (ACN) to 100% ACN. As a starting point, the method should run for 2–5 min at the initial injection solvent composition (95% H_2O/5% ACN) so that the peptide is equilibrated on the column before the gradient begins. Ramp up to 100% ACN over 15 min, followed by a 1–2-min holding step at 100% ACN with a 2-min ramp down to injection conditions before the next run begins. The flow rate will depend upon the chosen column, and the UV detector should monitor the wavelength at which the amide bond in the peptide backbone absorbs, 220 nm.

2. It is generally recommended to run a "blank" injection before purification where the mobile phase is injected to ensure that the column and HPLC equipment are clean.

3. Dissolve the peptide in the same solvent compositions present during injection (most often 95% H_2O with 5% ACN and 0.1% TFA). Shaking, vortexing, and sonication can all aid the dissolution process; however steps can be taken for difficult-to-solubilize peptides (see **Note 12**).

4. Inject a small amount of peptide (depending on the column, from less than a milligram to a few milligrams) to gain a better understanding of the shape of the HPLC peaks. In the case of purification problems, refer to **Note 13**.

5. Repeat HPLC runs until the peptide has been purified.

Combine the fractions from specific peaks across runs, remove the organic phase with rotary evaporation, and place into labeled tubes (most often 50 mL centrifuge tubes). Freeze the solutions inside the tube and lyophilize to produce a purified peptide powder. Mass spectroscopy, such as matrix-assisted laser desorption/ionization (MALDI) and electrospray ionization (ESI), can then be used to verify that the collected fractions have the correct (expected) mass. Peptides that will be used in tissue culture experiments or studies in regenerative medicine are typically sterilized before use. If the peptides can be dissolved in water without forming a gel they can filtered through a 0.22 μm filter to remove microbes. Peptides are also stable in most common organic solvents, such as ethanol, which is often used as a sterilization technique in laboratory experiments. It should be noted that neither of these techniques is effective against all pathogenic agents, including viruses or spores, and more robust sterilization techniques, such as ethylene oxide or gamma irradiation, can induce material degradation. Peptides sterilized through these techniques should be tested both for completeness of sterilization and for any chemical changes undergone during sterilization.

4   Notes

1. During synthesis, the C-terminus of the growing peptide chain is attached to the cross-linked polymer support, often referred to as the resin. A variety of resins are available but the two of the most common are Rink amide, which gives a C-terminal amide bond, and Wang resins, which often come preloaded with the first amino acid and form a C-terminal carboxylic acid when cleaved from the resin. Both of these resins are cleaved under strongly acidic conditions, such as trifluoroacetic acid above 90%. Other resins, such as the 2-chlorotrityl and Sieber resins, can be cleaved under mild acidic conditions (around 2% TFA) and give C-terminal carboxylic acids and amides, respectively.

2. HBTU is the most commonly used reagent to activate carboxylic acids in peptide synthesis, but there are a wide variety of other molecules available to promote amide bond formation. While HBTU is relatively inexpensive and prevents racemization (conversion of an enantiomerically pure mixture (where only one enantiomer is present) to a mixture of enantiomers), it is also a sensitizing agent and can induce allergic reactions

with repeated use. HCTU (2-(6-chloro-1H-benzotriazole-1-yl)-1,1,3,3-tetramethylaminium hexafluorophosphate) is similar to HBTU but reported to be less allergenic. For difficult couplings, HATU (2-(7-Aza-1H-benzotriazole-1-yl)-1,1,3,3-tetramethyluronium hexafluorophosphate) is often used as a more reactive coupling reagent.

3. The two most commonly used solvents for the coupling steps during peptide synthesis are dimethylformamide (DMF) and *N*-methylpyrrolidone (NMP). NMP has the advantage that it is more polar, which can improve coupling efficiencies and is less prone to degradation over time. However, DMF is often used because it is less expensive. In this chapter DMF and NMP are interchangeable for any coupling, deprotection, and washing steps.

4. Kaiser test solutions (also called ninhydrin test) are usually purchased as premade kits, but typically consist of the following three solutions:

    Solution A: 5 g ninhydrin dissolved in 100 mL ethanol.

    Solution B: 80 g liquified phenol in 20 mL ethanol.

    Solution C: 2 mL of 0.001 M KCN to 98 mL pyridine.

5. Solid support resins typically come at specified functionalizations that are given in millimoles (mmols) of functional groups per gram of resin. Functionalizations above 0.5 mmol/g are generally considered "high-loading" resins and below are considered "low loading." Having fewer functional groups in a low-loading resin is useful for improving coupling efficiency in long or difficult-to-synthesize peptides. Weigh out the appropriate quantity of resin, and place into the peptide synthesis vessel. For a peptide with a molecular weight of 1000 g/mol, a 1 mmol synthesis will have a theoretical yield of 1 g of peptide.

6. The N-terminus of the amino acid proline is unique in that the N-terminal amine is a secondary amine and not a primary amine. After coupling proline the Kaiser test following Fmoc deprotection typically does not turn dark purple, even if the Fmoc group has been removed. In the same vein the next amino acid coupling should be done with some care, as incomplete coupling is unlikely to give a positive Kaiser test. Once the next amino acid has been added the following Kaiser tests should proceed normally. For other non-proline amino acids, incomplete removal of the Fmoc-protecting group can be remedied using multiple incubations with 20% piperidine in DMF for longer periods of time. However, the factors that lead to incomplete peptide coupling, namely aggregation and reduced availability of the N-terminal amines to the activated peptide ester, are less problematic for Fmoc deprotection as the

piperidine molecule is significantly smaller than the activated amino acids and makes up a significant fraction of the solvent phase. In cases where there is difficulty removing the Fmoc group it is generally advisable to cleave a small amount of resin and check the mass of the growing peptide to ensure that unwanted reactions have not stopped the normal synthesis process.

7. When weighing out amino acids it is important to bring all reagents to room temperature before weighing them out to prevent condensation inside the containers. HBTU is allergenic and should be weighed out in a chemical hood, and it is often easiest to weigh out all activation agents (HBTU, etc.) and amino acids into vials or centrifuge tubes before starting the synthesis and then using them as needed. The tubes can be refrigerated until needed; however powders are generally stable at room temperature over the timescales of a peptide synthesis. As an example, for a 0.25 mmol synthesis (which has a theoretical yield of 250 mg of a 1000 MW peptide), four equivalents of the amino acid is equal to 4×0.25 mmol or 1 mmol of the amino acid. For Fmoc-glycine, with a molecular weight of 297 g/mol, this is 297 mg. HBTU, with a molecular weight of 397 g/mol, is added at 3.95 equivalents, which is 392 mg for a 0.25 mmol synthesis. A sixfold excess of DIPEA in this case is 260 μL. A fourfold excess of amino is typical for coupling steps as it helps to ensure that the reaction goes to completion, but lower excesses can be used for expensive or precious coupling steps. The number of equivalents of HBTU needs to be less than the amino acid to prevent guanidinylation of the N-terminus of the peptide, and DIPEA is typically used in 1.5-fold excess to the amino acid.

8. Incomplete couplings will yield Kaiser tests with varying degrees of purple resin or solution. Coupling steps can be repeated as many times as needed before the Fmoc group is deprotected, and other steps can be taken to improve the yield of problematic couplings (*see* **Note 9**). If some free amines are still present after a coupling another option is to "cap" them with a group which renders the amines unreactive in subsequent couplings. The benefit of capping is that it terminates the chain and typically makes the short caped peptide easier to remove during purification than full-length peptides with an amino acid deletion. Anhydrides react with amines and capping is most often done with acetic anhydride, in a 10:2.5:100 (v/v/v) acetic anhydride:DIPEA:DMF solution. This is repeated twice for 5 min. It is worth emphasizing that the resin and interior of the synthesis vessel and cap should be thoroughly washed after capping, as the next step is typically Fmoc deprotection with piperidine and free acetic anhydride will

react with the newly deprotected amines. Once the final amino acid has been added, the last Fmoc group is removed with a 20% piperidine in DMF solution. The resin is then washed twice with DMF and three times with DCM before cleavage, unless modification of the N-terminus is desired (*see* **Note 10**).

9. Self-assembling peptides are designed to aggregate, and this can cause problems during synthesis. There are two different routes to reducing aggregation: (1) modifying conditions inside the shaker vessel and (2) adjusting the structure of amino acids to reduce intermolecular hydrogen bonding. A flowchart going through the troubleshooting of difficult couplings is presented in Fig. 4 and explained in more detail below.

(a) Reducing aggregation during synthesis.

The amide coupling reaction is more efficient with more polar solvents like DMF and *N*-methylpyrrolidone (NMP) than DCM. However, most solid support resins tend to swell more in DCM, which spaces out the peptide chains and reduces aggregation. A common mixture uses a 1:1:1 mixture of DMF:NMP:DCM during difficult coupling steps. Dimethylsulfoxide (DMSO) has also been added to reaction mixtures to reduce aggregation during synthesis. More efficient coupling reagents, such as HATU, can be used during difficult steps. Finally, other chemicals can be added to the reaction mixture to break up aggregation. Surfactants, such as Triton X-100, can be used to reduce hydrophobic collapse in polar solvents, although it is extremely important that the surfactant does not have primary amines or carboxylic acids and thus does not take part in the reaction. Other steps to reduce hydrogen bonding include using chemicals such as lithium bromide, which is soluble in DMF and can be added during synthesis.

(b) Designing peptides for reduced aggregation.

Secondary structure and aggregation in peptides and proteins are primarily driven by hydrogen bonding between different amide bonds. There are a variety of ways in which this can be inhibited or reduced. The first is to physically increase the distance between peptide chains, which can be achieved through modifying the choice of resin. Switching to a resin with a lower peptide loading range will reduce aggregation and increase yields of self-assembling peptides. Another option is to choose a resin with an increased ability to swell, often through the incorporation of poly(ethylene glycol) (PEG) into the resin. It is also possible to modify the peptide sequence to include amino acids, such as proline, which disrupt aggregation. The benefit of incorporating prolines in a sequence is that it is a natural amino acid and does not require the purchase of special

Difficult peptide synthesis flow chart

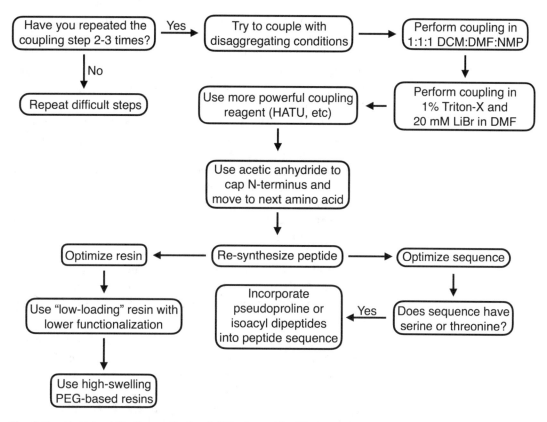

Fig. 4 Flowchart to aid in the synthesis of difficult peptides. During synthesis the first steps are to repeat the difficult couplings (before removing the Fmoc-protecting group), following by using solvent mixtures and additives designed to reduce aggregation during synthesis. Other actions that can be taken during the synthesis include using more powerful coupling reagents or "capping" the N-terminus with a nonreactive acetyl group which will prevent further additions to the free N-termini and make removal of truncated peptides easier during purification. It is also possible to improve synthesis through selection of solid-phase resin and optimization of the peptide sequence. Using lower loading or higher swelling resins typically increases the distance between peptide chains during synthesis which reduces aggregation. Furthermore, sequences which contain serine or threonine can utilize pseudoproline or isoacyl dipeptide residues which reduce aggregation during synthesis but revert back to normal amino acids either during peptide cleavage or at basic pHs

amino acids, or more complicated synthetic chemistries. However, since most desired peptide sequences cannot be easily modified, chemists have developed alternate ways to break up aggregation. One way is to mimic the disruptive five-membered ring of proline by creating "pseudoproline" amino acids in which the side chain is coupled to the backbone in a way which reverts back to the native amino acid under peptide cleavage conditions. These pseudoproline amino acids are typically sold as dipeptides, since the pseudoproline modification renders the amine less reactive. These are available for serine, threonine, or cysteine. The pseudoproline linkages revert to the standard amino acid in TFA concentrations above 50%, so it is possible to cleave the peptide from resins that are sensitive to weakly acidic conditions (such as 2-chlorotrityl and Sieber resins), and purify the peptide with protecting groups and pseudoprolines still present. In these cases the peptides are often not soluble in pure water, and will need to be dissolved in acetonitrile/water mixtures. Another synthetic route is to use "iso-acyl dipeptides" which break up aggregation but revert to normal peptide sequences under slightly basic conditions after cleavage from the resin. These have an added benefit in that this shift back to standard peptide linkages does not occur during cleavage, as is the case with pseudoprolines, which can make purification easier for peptides that aggregate (if the peptide is purified under acidic conditions). For peptides that do have pseudoprolines, it is possible to cleave them under mild acidic conditions with a Sieber or 2-chlorotrityl resin, purify them with the pseudoproline still present on the peptide, and then cleave the final product in a typical cleavage cocktail.

10. An important consideration before cleavage is whether or not a free N-terminal amine is desired on the peptide. If a free amine on the N-terminus is undesirable it can be capped with the same acetic anhydride solution as the previously described capping reaction, which will acetylate the N-terminal amine. The N-terminus can also be modified with functionalizations found in proteins, such as palmitoyl and myristoyl group. Hydrocarbon chains on the end of peptide sequences are a popular motif to induce self-assembly, and the basis for a class of self-assembling "peptide amphiphiles" [2]. Aliphatic acids, such as myristic acid, oleic acid, or palmitic acid, can be added to the peptide through the same coupling procedure as an amino acid, with activation of the carboxylic acid in DMF which is then added to the resin-bound peptide. Another group of self-assembling peptides uses the Fmoc group present on the final amino acid to help induce self-assembly. In this case the final Fmoc group is left on the resin and the peptide is cleaved with the Fmoc group still present [6]. Once the coupling step is done the resin is washed and cleaved.

11. Most cleavages are done with 95% TFA, 2.5% TIPS, and 2.5% H_2O. However, for peptides containing cysteine, methionine, or tryptophan, this can lead to unwanted side reactions or incomplete protecting group removal. This is normally ameliorated through the addition of either ethanedithiol (EDT) or dithiothreitol (DTT). EDT has the benefit that it is significantly cheaper than DTT; however it has a pungent odor that

is difficult to completely remove from the peptide, and typically requires that glassware it comes in contact with is submerged in a base bath. DTT is a white, powdery solid and can be added at 2.5 wt% in the cleavage solution (TFA/DTT/TIPS/H$_2$O of 92.5/2.5/2.5/2.5).

12. A variety of steps can be used to improve the solubility of peptides. Typically the first steps include sonication at raised temperatures (50 °C), or addition of additional acetonitrile for peptides which are especially hydrophobic. It should be noted that if acetonitrile is added to the peptide to aid dissolution, the composition of the mobile phase running in the HPLC during injection should be increased to match that within the peptide solution. In cases where this is not sufficient, the peptide can be lyophilized and dissolved in TFA to break up any aggregates. The TFA is then removed and the peptide is dissolved in the mobile phase. Further steps include dissolving the peptide in good peptide solvents (such as DMSO) and then adding mobile phase; however this should be done with caution as the peptides can become stuck on the column and damage it. For peptides which are negatively charged (have more carboxylic acids than positively charged groups), it is important to note that there is typically some residual TFA left after the cleavage steps, so adding base (including NH$_4$OH, NaOH, or KOH) will probably be necessary to raise the pH above 7 before the peptide will dissolve in water.

13. Purification problems typically manifest themselves in three different ways. The first is that the peptide will not stick to the HPLC column and elutes within the first few minutes, in the "breakthrough" peak. To improve peptide retention, chaotropic agents, such as 20 mM ammonium formate, can be added to the aqueous phase. Some HPLCs have column heaters which can help both retention and peak sharpness. Another method to improve peptide retention is to include more groups which will interact with the HPLC column. This can be done by including more hydrophobic amino acids, but it can also be done by cleaving the peptide under mild acidic conditions using a Sieber or 2-chlorotrityl resin, as many protecting groups are hydrophobic. Once the purified, protected peptide has been dried, it can be removed with the proper cleavage cocktail.

Ideally a peptide forms a single, sharp peak on HPLC allowing for it to be easily separated from impure fractions. However, for peaks that are poorly defined it is possible to try and improve their shape. The inclusion of chaotropic agents (20 mM ammonium formate in the aqueous phase), heating the column, and sonication of the solution prior to inject can all help. The HPLC method can be changed to slow down the gradient during the elution times. It is important to note that the solvent can take several minutes to travel from the pumps to the column, depending on the flow rate, injection loop size, and column volume, so the composition inside the column is typically the composition that left pump a few minutes earlier. Slower gradients can help space out peaks and improve purity, at the cost of longer runs.

Finally, some peptides have a strong affinity for the column and do not fully come off. This is seen if the pressure in the HPLC increases after each run. This is most common with peptides that have significant hydrophobicity, especially alkyl tails. Raising the column temperature, and including chaotropic agents while injecting solvents such as isopropanol, can help to remove bound peptides.

References

1. Webber MJ, Appel EA, Meijer EW, Langer R (2016) Supramolecular biomaterials. Nat Mater 15:13–26
2. Hartgerink JD, Beniash E, Stupp SI (2001) Self-assembly and mineralization of peptide-amphiphile nanofibers. Science 294:1684–1688
3. Palomo JM (2014) Solid-phase peptide synthesis: an overview focused on the preparation of biologically relevant peptides. RSC Adv 4:32658–32672
4. Coin I, Beyermann M, Bienert M (2007) Solid-phase peptide synthesis: from standard procedures to the synthesis of difficult sequences. Nat Protoc 2:3247–3256
5. Cui H, Pashuck ET, Velichko YS et al (2010) Spontaneous and x-ray-triggered crystallization at long range in self-assembling filament networks. Science 327:555–559
6. Jayawarna V, Ali M, Jowitt TA et al (2006) Nanostructured hydrogels for three-dimensional cell culture through self-assembly of fluorenylmethoxycarbonyl–dipeptides. Adv Mater Weinheim 18:611–614

Chapter 15

H$_2$S Delivery from Aromatic Peptide Amphiphile Hydrogels

Kuljeet Kaur, Yun Qian, and John B. Matson

Abstract

Hydrogels are materials composed mostly of water that have found use in a wide variety of applications, including tissue engineering and regenerative medicine. Aromatic peptide amphiphiles can be designed to self-assemble in aqueous solution into one-dimensional aggregates that entangle to form hydrogels with very high water content (>99 wt. %). Here, we describe the synthesis of an aromatic peptide amphiphile designed to release hydrogen sulfide (H$_2$S), a vital biological signaling gas with significant therapeutic potential. Peptide synthesis, purification, aliquotting, and procedures for measuring H$_2$S release are detailed.

Key words Solid-phase peptide synthesis, Gasotransmitters, Aromatic peptides, Hydrogen sulfide (H$_2$S), Self-assembly

1 Introduction

H$_2$S has emerged over the past 20 years as a biological signaling agent that plays a role in many physiological processes. It joins nitric oxide and carbon monoxide on the list of endogenous signaling gases called gasotransmitters [1]. H$_2$S is produced in the body primarily by two enzymes: cystathionine gamma-lyase (CSE) and cystathionine beta-synthase (CBS) [2]. Through the action of these enzymes, H$_2$S plays a role in many physiological processes including angiogenesis [3], vasodilation [4], inflammation [5], and neuromodulation [6]. It can also protect cells from harsh environments by suppressing oxidative stress [7]. However, its progress toward clinical applications has been hindered due to an inability to limit its release to a desired location at a controllable rate. Current research in the field of H$_2$S is focused on developing materials capable of achieving these goals.

One method to control the release of drugs is to incorporate them into hydrogels that can be implanted or injected at a site of action. Controlled release is especially important for H$_2$S because of its volatile nature and fast diffusion and reactivity in the body.

Kanika Chawla (ed.), *Biomaterials for Tissue Engineering: Methods and Protocols*, Methods in Molecular Biology, vol. 1758, https://doi.org/10.1007/978-1-4939-7741-3_15, © Springer Science+Business Media, LLC, part of Springer Nature 2018

Hydrogels are networked structures that are capable of organizing a large volume of water. There are two basic types of hydrogels: those that form networks via chemical cross-linking, known as chemical hydrogels, and those that form networks based on self-assembled structures, known as physical hydrogels. Various peptide designs based on hydrogen bonding, hydrophobic interactions, ionic interactions, and/or aromatic stacking are capable of forming physical hydrogels [8]. Peptide-based physical hydrogels have been used widely in tissue engineering and regenerative medicine for applications in cardiac tissue repair [9], wound healing [10], bone regeneration [11], drug delivery [12], and many others.

A wide variety of physical hydrogels based on aromatic peptide amphiphiles have been prepared, indicating the robustness and versatility of this design motif [13]. Aromaticity in the peptides can be included via modification of the N-terminus (for example by the Fmoc group, a protecting group used widely in peptide synthesis) or by incorporation of aromatic amino acids into the peptide backbone. Here we describe the synthesis of an aromatic peptide amphiphile with amino acid sequence IAVEEE [depicted in (**D**) Fig. 2] [14] in which the N-terminus is modified with an *S*-aroylthiooxime (SATO) group. SATO functional groups release H_2S gas in response to thiols (e.g., cysteine or glutathione) [15]. The aromatic peptide amphiphile (**D**) is synthesized by solid-phase peptide synthesis (SPPS) with the addition of 4-formylbenzoic acid to the N-terminus. After cleavage from the resin, *S*-benzoylthiohydroxylamine (SBTHA) is added to form the SATO functional group. The resulting peptide forms a robust hydrogel upon addition of $CaCl_2$ and is capable of releasing H_2S, in a controlled manner, when triggered with cysteine.

2   Materials

2.1   Reagents for Peptide Synthesis and Hydrogel Formation

1. 4-(2′,4′-Dimethoxyphenyl-Fmoc-aminomethyl)phenoxyacetamido-methylbenzhydryl amine resin (Rink-amide MBHA resin).

2. Dichloromethane (DCM).

3. *N*,*N*-Dimethylformamide (DMF).

4. Trifluoroacetic acid (TFA).

5. Ethyl acetate (EtOAc).

6. Methanol (MeOH).

7. Ethanol (EtOH).

8. Diethyl ether (Et_2O).

9. Toluene.

10. Hexanes.

11. Triisopropylsilane (TIPS).

12. Dimethylsulfoxide (DMSO).

13. Water.

14. Fmoc-Ile-OH.

15. Fmoc-Ala-OH.

16. Fmoc-Val-OH.

17. Fmoc-Glu(OtBu)-OH.

18. *N*,*N*,*N'*,*N'*-Tetramethyl-*O*-(1H-benzotriazol-1-yl)uronium hexafluorophosphate (HBTU).

19. *N*,*N*-Diisopropylethylamine (DIEA).

20. 4-Methylpiperidine.

21. 1,8-Diazabicyclo[5.4.0]undec-7-ene (DBU).

22. Potassium cyanide (KCN).

23. Ninhydrin.

24. Pyridine.

25. Phenol.

26. Ammonium hydroxide (30% solution in water).

27. 4-Formylbenzoic acid.

28. *p*-Toluenesulfonic acid.

29. Ethylene glycol.

30. Potassium hydroxide (KOH).

31. Thiobenzoic acid.

32. Hydroxylamine-*O*-sulfonic acid.

33. Silica gel.

34. Potassium carbonate (K$_2$CO$_3$).

35. Phosphate buffer (0.1 M, pH 7.4).

36. Sodium sulfide (Na$_2$S).

37. Calcium chloride (CaCl$_2$).

38. Ethylenediaminetetraacetic acid (EDTA).

39. Cysteine.

Fmoc-protected amino acids and RINK-amide-MBHA resin should be stored at <10 °C (*see* **Note 1**). Follow all waste disposal protocols (*see* **Note 2**).

2.2 Glassware and Related Small Equipment

1. Shaker vessel (a cylindrical glass vessel containing a screw-cap inlet at the top and a coarse glass frit at the bottom with a drain stem equipped with a stopcock).

2. Cap for shaker vessel including PTFE liner.

3. Stir bar.

4. Dean-Stark trap.

5. Condenser.

6. Round-bottom flask (100 mL).

7. Erlenmeyer flask (250 mL).

8. Glass funnel with coarse frit (60 mL).

9. Glass funnel with fine or medium frit (30 mL).

10. Graduated cylinders (10, 250, 500 mL).

11. Amber bottles (250, 25 mL).

12. 100 μL glass syringe.

13. Vacuum chamber.

2.3 Consumables

1. TLC plates.

2. Microcentrifuge tubes (1.5 mL).

3. Pasteur pipettes.

4. Glass test tubes (4 mL).

5. Plastic centrifuge tubes (50 mL).

6. Glass vials (20 mL).

7. Glass vials (4 mL).

8. Kimwipes.

9. Rubber bands.

10. Glass capillary tubes (open ended).

11. 3 Å molecular sieves.

12. Gas chromatography vials (1.5 mL).

13. Plastic syringe (10 mL).

14. Syringe filters (25 mm with 0.45 μm PTFE membrane).

15. pH paper (0–14).

2.4 Solutions

1. **Deprotection solution**: *Composition*: 2% 4-methylpiperidine and 2% 1,8-diazabicyclo[5.4.0]undec-7-ene (DBU) (v/v) in *N,N*-dimethyl formamide (DMF). In order to prepare 500 mL of the solution: Measure 10 mL of 4-methylpiperidine into a small graduated cylinder and transfer it to a 500 mL graduated cylinder. Using another small graduated cylinder, measure 10 mL DBU and transfer it to the same 500 mL graduated cylinder containing 4-methylpiperidine. Fill up to 500 mL with DMF (*see* **Note 3**). Store this solution at room temperature in an amber bottle (*see* **Note 4**).

2. **Cleavage solution**: *Composition:* 2.5% water and 2.5% triisopropylsilane in trifluoroacetic acid (TFA). In order to prepare 250 mL of the solution: Measure 6.25 mL of water using a graduated cylinder or syringe; pour it into a 250 mL graduated cylinder. Add 6.25 mL of triisopropylsilane using a graduated

cylinder or syringe. Make up the solution to 250 mL with TFA (*see* **Note 5**). Store this solution at room temperature in an amber bottle (*see* **Note 6**).

3. **HPLC purification, mobile-phase solvents:** *Composition*: 0.1% ammonium hydroxide (NH$_4$OH) in water and 0.1% ammonium hydroxide (NH$_4$OH) in acetonitrile. In order to prepare 4 L of each solution: To a 4 L bottle of acetonitrile or water, add 4 mL of ammonium hydroxide (30% in water). Store this solution at room temperature in an amber bottle indefinitely.

2.5   Equipment

1. RP-HPLC: Agilent Technologies 1260 Infinity HPLC system equipped with a fraction collector using an Agilent PLRP-S column (100 Å particle size, 25 × 150 mm) and monitoring UV absorbance at 220 nm.

2. H$_2$S probe: 2 mm H$_2$S selective microelectrode, World Precision Instruments.

3. Lyophilizer: 6 Lt. Cascade Freeze dryer, 103–127 V, Labconco.

4. Shaker: Fisher Scientific fixed-speed wrist motion shaker, 400 oscillations/min.

5. Mass spectrometer: Advion ExpressIon Compact Mass Spectrometer, ESI probe with negative-ion source settings.

6. Schlenk line equipped with inert gas and vacuum manifolds.

7. Glove box.

8. Rotary evaporator.

9. Stir plate.

10. Handheld UV lamp.

11. Scissors.

12. Thumbtack.

13. Cardboard insert from a cryo-storage box.

14. Freezer (−20 °C).

3   Methods

Carry out all steps at room temperature and inside chemical fume hoods unless otherwise specified.

3.1   Synthesis of Aldehyde-Terminated Peptide Amphiphile (Fig. 1)

3.1.1   Synthesis of the Peptide on Resin

1. Add 1 mmol (*see* **Note 7**) of RINK-amide-MBHA resin (typical resin loading is 0.3–0.8 mmol/g) to a 100 mL peptide synthesis shaker vessel.

2. Add 15 mL of DMF to the shaker vessel and shake for 15–20 min to swell the resin.

3. Deprotect the resin (remove the Fmoc group) by adding 10 mL of deprotection solution to the shaker vessel and shaking

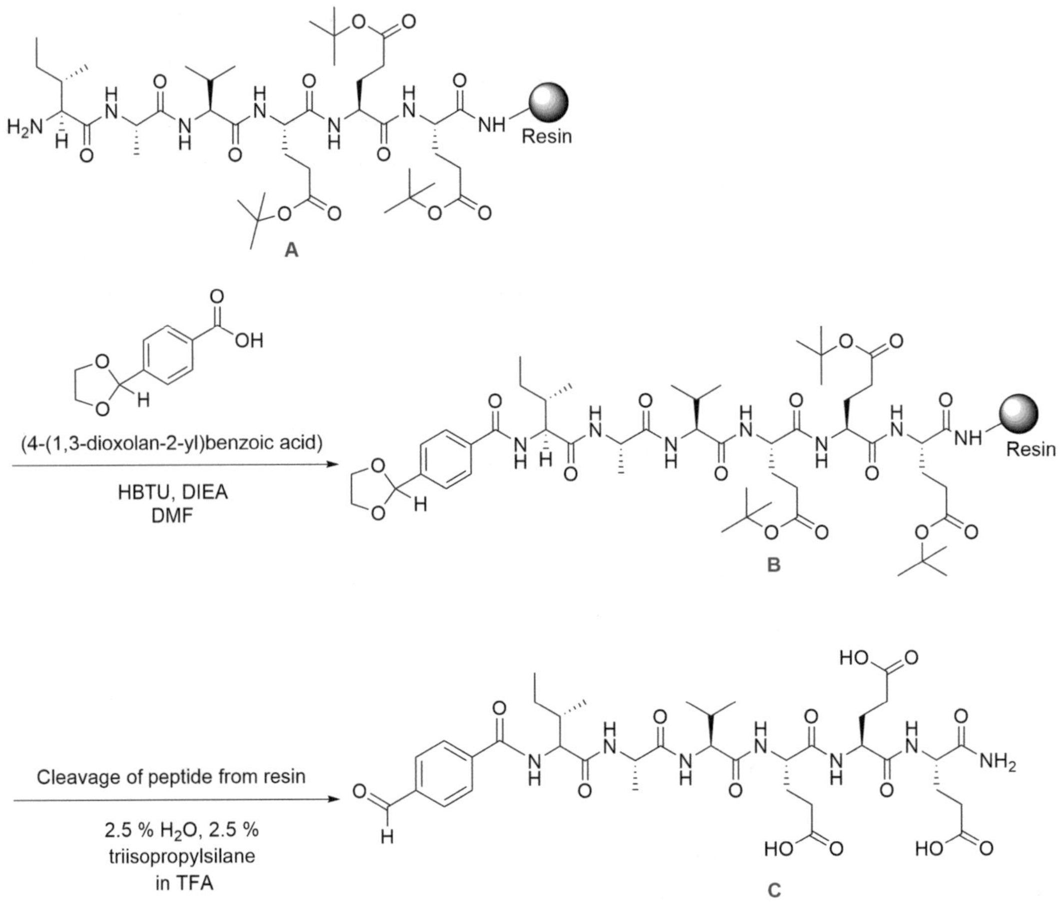

Fig. 1 Synthetic scheme showing coupling of Fmoc-deprotected peptide (**A**) to 4-(1,3-dioxolan-2-yl)benzoic acid to give an aldehyde-terminated peptide amphiphile attached to the resin (**B**) and chemical structure of the final aldehyde-terminated peptide amphiphile (**C**) after cleavage from the resin

it for 5 min on a wrist-action shaker. Drain the deprotection solution and wash the resin with DMF (2 × 15 mL). Add another 10 mL of the deprotection solution to the resin and shake the shaker vessel for another 5 min. Again, drain the vessel and wash the resin with DMF (2 × 15 mL) and dichloromethane (DCM) (2 × 15 mL).

4. Remove several resin beads (approximately 10–15) from the vessel using a Pasteur pipet, and put them inside a small glass test tube. Confirm that the deprotection reaction was successful using the Kaiser test (*see* **Note 8**). If the color of the resin beads is purple, then proceed to next step; otherwise, remake the deprotection solution and repeat **step 3**.

5. To prepare the "coupling cocktail," mass 4.0 mmol (1.70 g) of Fmoc-Glu(OtBu)-OH (or the desired Fmoc-protected amino acid) and 3.9 mmol (1.48 g) of 2-(1 H-benzotriazole-1-yl)-1,1,3,3-tetramethyluroniumhexafluorophosphate   (HBTU)

into a plastic 50 mL centrifuge tube or a 20 mL glass vial (*see* **Note 9**). This mixture of solids can be prepared before use and stored for several days at room temperature. To the Fmoc-amino acid/HBTU mixture, add 6 mmol (1.07 mL) of *N,N*-diisopropylethylamine (DIEA) and 10–15 mL of DMF (*see* **Note 10**). Shake the contents of the centrifuge tube until homogeneous. A vortexer may be used here. Pour this coupling cocktail into the peptide synthesis vessel and shake it for 2.5–3.0 h.

Drain the coupling cocktail from the peptide synthesis vessel and wash the resin beads thoroughly with DMF (3 × 15 mL) and then DCM (3 × 15 mL). The success of coupling can be checked using the Kaiser test in the same manner as explained in **step 4**. A successful coupling is indicated by a light yellow solution and no blue/purple color in the resin beads (*see* **Note 8**).

6. Repeat the deprotection and the coupling steps for each additional amino acid. A Kaiser test after each coupling and each deprotection step is recommended. If the Kaiser test indicates that the step was unsuccessful, then repeat the step.

7. Once the sequence is complete (IAVEEE in this case), carry out one final deprotection step to remove the Fmoc-protecting group and expose a free amine.

8. To add the arylaldehyde unit, transfer 4 mmol (0.776 g) of 4-(1,3-dioxolan-2-yl)benzoic acid (*see* **Note 11**) into a plastic 50 mL centrifuge tube (*see* **Note 12**). Add 3.9 mmol (1.48 g) HBTU, 6 mmol (1.07 mL) DIEA, and 10–15 mL DMF as listed above in **step 5**. Shake or vortex until homogeneous, and then add the solution to the peptide synthesis vessel. Shake for 2.5–3.0 h.

9. Drain the synthesis vessel and wash the resin with DMF (1 × 15 mL).

10. Repeat **steps 8** and **9** to ensure complete coupling.

11. Check the coupling using Kaiser test. Repeat once more if a blue/purple color is observed.

3.2 Cleavage of the Peptide from the Resin

1. Add about 10 mL of the cleavage solution to the peptide synthesis vessel. Shake it for 1–2 h (*see* **Note 13**). The solution may turn red during this step.

2. Drain the cleavage solution into a 100 mL round-bottom flask, and wash the resin with DCM (2 × 10 mL). Combine the washings with the peptide solution in the round-bottom flask.

3. Concentrate the entire solution in vacuo by rotary evaporation until about 1 mL of solution is left.

4. Add approximately 25–30 mL of cold diethyl ether to the round-bottom flask to precipitate out the peptide. Filter (*see* **Note 14**) and wash the white or off-white precipitates with diethyl ether. Collect the sample in a vial or 50 mL centrifuge tube and dry it under vacuum to afford the crude product. Crude yields typically range from 80 to 140%, which includes residual water and TFA that do not need to be removed entirely at this stage.

3.3 Purification via RP-HPLC

1. To the crude peptide product, add deionized water (approximately 1 mL water per 10 mg product). Next, add 30% NH_4OH solution dropwise to the peptide solution to solubilize it. Sonicate the solution until the peptide is fully solubilized. Filter the peptide solution using a 10 mL syringe equipped with a 25 mm syringe filter with 0.45 μm PTFE membrane.

2. Purify the peptide by preparative-scale HPLC using a gradient of 2–50% acetonitrile in water (both having 0.1% NH_4OH by volume) over 30–35 min at a flow rate of 10 mL/min.

3. Analyze fractions collected after purification by mass spectrometry (*see* **Note 15**).

4. Combine pure product-containing fractions into a round-bottom flask of at least twice the volume of the solution. Concentrate the solution via rotary evaporation to remove most of the acetonitrile. Transfer the aqueous solution into one or more tared 50 mL conical centrifuge tubes. Do not add more than 40 mL to any tube.

5. Freeze the solutions in the 50 mL centrifuge tubes with their caps on in liquid nitrogen for 15 min. Once thoroughly frozen, remove each cap and replace it with a Kimwipe secured with a rubber band. Transfer the frozen centrifuge tubes onto a lyophilizer for 1–2 days to remove the water.

3.4 Functionalization to form SATO-Peptide (Fig. 2)

3.4.1 Synthesis

1. Transfer 200 mg of the lyophilized peptide powder carefully into a tared 4 mL glass vial (*see* **Note 16**).

2. Dissolve the peptide in 200 μL dry DMSO (*see* **Note 17**), and add 182 mg (5 equiv.) of S-benzoylthiohydroxylamine (SBTHA) (*see* **Note 18**) to it. Shake until both reactants are fully dissolved.

3. Add 54 μL (3 equiv.) of TFA to the above reaction mixture using a glass syringe. Add activated 3 Å molecular sieves to the vial to absorb the water that forms during the reaction.

4. Flush the vial with nitrogen gas, seal it, and let it stand. Reaction progress can be monitored by mass spectrometry. To do this, remove approximately 20 μL of the reaction mixture from the reaction vial using a glass capillary or a Pasteur pipet.

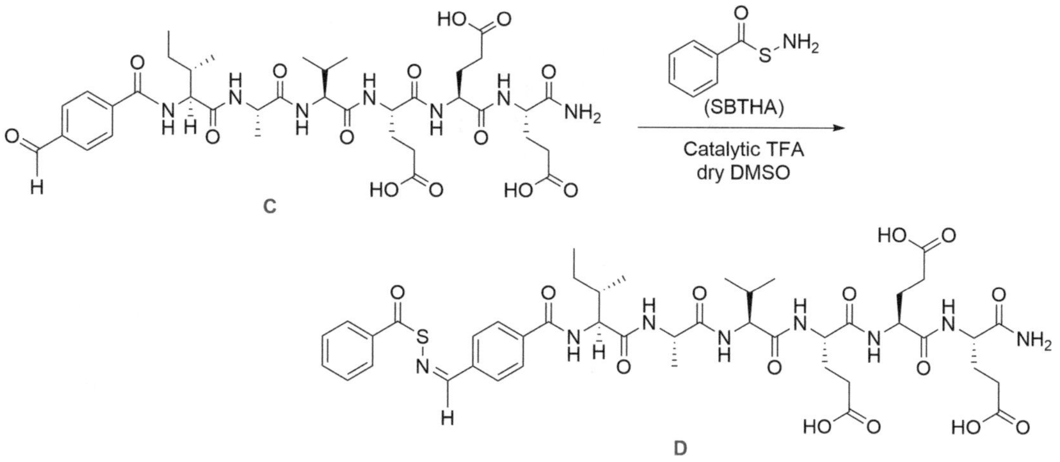

Fig. 2 Reaction of aldehyde-terminated peptide amphiphile (**C**) with *S*-benzoylthiohydroxylamine (SBTHA) to give H₂S-releasing aromatic peptide amphiphile (**D**)

Dilute this aliquot with 0.5 mL of methanol and inject it into mass spectrometer. The reaction is usually complete in less than an hour; if not, repeat the mass spectrometry monitoring until the reaction is complete.

5. When complete, pour the reaction mixture into a 250 mL Erlenmeyer flask containing 100 mL DCM. Wash the molecular sieves with an additional 15 mL DCM. The peptide product will appear as a floating white mass in DCM.

6. Collect the peptide residue using vacuum filtration.

7. Transfer the residue into a tared 20 mL glass vial or a 50 mL conical centrifuge tube.

8. Place the vial in a vacuum chamber and apply vacuum to remove excess solvent.

3.5 Purification

1. Add about 100 mg of crude peptide **D** into a 50 mL plastic centrifuge tube and add 10 mL deionized water to it. Adjust the pH to 7.5 by carefully adding K₂CO₃ solution (10 mg/mL in deionized water) dropwise to the peptide solution. Measure the pH with pH paper or a calibrated pH meter (*see* **Note 19**).

2. Purify the peptide following the steps used to purify crude peptide **C** above, omitting 0.1% NH₄OH in each mobile phase.

3. Store the purified peptide as a lyophilized powder in a 50 mL conical centrifuge tube in the freezer.

3.6 Aliquotting the Peptides

Aliquot the peptide in masses appropriate for various analyses in order to minimize waste and repeated removal from the freezer. This is vital for maintaining a consistent peptide sample concentration for each type of analysis. If the peptide is instead repeatedly removed from its storage container, then the water content in each sample can change because the peptide can absorb water when exposed to the atmosphere. Water can also promote hydrolysis of the SATO functional group, degrading the peptide.

Decide on the size and number of aliquots first. For this example, ten 1 mg aliquots of peptide will be prepared.

1. Use 1.5 mL microcentrifuge tubes for the aliquots. Obtain 20 tubes, and cut off the caps from 10 of them using scissors. Using a thumbtack, a needle, or a similar sharp object, make a hole through the top of each of the ten removed caps.

2. Dissolve 10 mg of pure peptide in 1 mL deionized water, and transfer 0.1 mL of the peptide solution to each of the whole microcentrifuge tubes.

3. Cover each of the microcentrifuge tubes containing the peptide solution with a pierced cap as prepared in **step 1**.

4. Create a small tray to hold the microcentrifuge tubes using the cardboard insert from a cryo-storage box. Put the ten capped microcentrifuge tubes in the box and freeze in liquid nitrogen for 10 min. Lyophilize the aliquots until dry.

5. Once dry, remove the box of tubes from the lyophilizer. Carefully tap down any of the lyophilized powder that has stuck to the cap. If any powder cannot be tapped back into the base of the tube, then remove this tube from the set of aliquots and discard it. Remove the pierced caps and seal the microcentrifuge tubes with attached (intact) caps.

6. Store the aliquots in the freezer. When needed for analysis, allow them to warm to room temperature before use.

3.7 Gel Formation

1. Dissolve 10 mg of the pure lyophilized peptide in 1 mL deionized water. We observe that the lyophilized peptide is generally soluble; if it is not, then carefully add K_2CO_3 solution (10 mg/mL in deionized water) until the pH is 7.5. Confirm the pH using pH paper or a calibrated pH meter.

2. Prepare 10 mL of a $CaCl_2$ solution (20 mM) in deionized water.

3. Add 250 µL of the peptide solution prepared in **step 1** into a small glass vial (for example, a 1.5 mL gas chromatography vial). Add 25 µL of the $CaCl_2$ solution to the peptide solution.

4. Formation of a self-supporting gel should be observed in less than 1 min. It can be confirmed by turning the glass vial upside-down (Fig. 3).

3.8 H_2S Release Experiments

3.8.1 Calibration of the Probe

1. Prepare 25 mL of an EDTA solution (154 µM) in deionized water and purge by vigorously bubbling with nitrogen gas for 20 min.

2. To a 20 mL glass vial, add 3.85 mg of anhydrous sodium sulfide (Na_2S) (*see* **Note 20**) followed by 10 mL of the EDTA

Fig. 3 Self-supporting gel in an upside-down vial

solution to get the final 5 mM concentration. This is the H$_2$S solution used for calibration.

3. Add a small stir bar to a 20 mL glass vial containing 10 mL of 1× PBS buffer (pH 7.4) and place it on a stir plate.

4. Immerse the H$_2$S probe in the PBS solution and allow the background current to stabilize (this usually takes approximately 5 min).

5. Using a glass syringe or manual pipette, sequentially add five aliquots (20, 40, 60, 80, 100 µL, in order to construct a reliable calibration curve) of H$_2$S solution into the vial, and observe the spike in current with each addition. Let the current reach a plateau in between each addition, which should take about 5–10 min.

6. Construct a linear calibration curve of concentration vs. current after calculating the total concentration of H$_2$S in solution after addition of each aliquot.

3.8.2 Release From Peptide D in Solution

1. Prepare 10 mL of a stock solution of cysteine (20 mM) in PBS buffer (pH 7.4).

2. Prepare a solution of 100 µM peptide **D** in PBS buffer as follows. Add 1 mg of peptide **D** to a 1.5 mL plastic centrifuge tube and add 1 mL PBS buffer solution. Dilute the 1 mg/mL solution to 100 µM by adding 9.17 mL PBS buffer solution to 0.83 mL of the 1 mg/mL stock solution in a separate vial.

3. Add 3.8 mL of the 100 µM peptide **D** solution into a 20 mL glass vial. In order to maintain consistency in conditions with H_2S release from the solution, do not stir the solution.

4. Submerge the H_2S selective probe into the peptide **D** solution and allow it to equilibrate over 3–4 min.

5. Add 200 µL of the cysteine solution into the above vial to get a final concentration of 1 mM cysteine.

6. Monitor the current without stirring the vial during analysis.

7. The plot of current vs. time that the instrument produces can be exported and converted into a plot of H_2S concentration vs. time using the calibration curve obtained earlier.

8. Because the solution is not stirred, the data can be noisy. If needed, smooth the data using a Savitzky-Golay function (Igor software) or other smoothing functions.

3.8.3 Release From Peptide D Gel

1. Prepare a 1 wt. % solution of peptide **D** by dissolving 1 mg of peptide in 100 µL deionized water.

2. Add 40 µL of the peptide **D** solution to a 4 mL glass vial or any similar glass vial available.

3. Add 4 µL of a 20 mM of $CaCl_2$ solution in deionized water to induce gelation.

4. Add 3.9 mL of PBS buffer solution to the above vial, and submerge the microelectrode into the solution without disturbing the gel. Allow the current to stabilize (this usually takes ~5 min).

5. Add 20 µL of a 200 mM cysteine solution in deionized water to the same vial to get a final concentration of 1 mM cysteine. Proceed by following **steps 6–8** above.

4   Notes

1. Avoid allowing both the RINK-amide-MBHA resin and the Fmoc-amino acids to absorb water. Store bottles in the refrigerator, and let them sit at room temperature for 15–20 min before opening.

2. Wear latex or nitrile gloves and other proper personal protective equipment while handling and disposing of all chemicals, as most of them are toxic. Latex gloves are preferable because they protect against exposure to DMF better than nitrile gloves. Centrifuge tubes used to prepare the coupling cocktail (activation of amino acids) and the glass tubes used for the Kaiser test should be rinsed with acetone before disposal in accordance with all regulations.

3. An alternative deprotection solution should be used if the amino acids asparagine or aspartic acid are included in the peptide because DBU can induce aspartimide formation. To avoid this side reaction use 30% of 4-methylpiperidine (v/v) in DMF.

4. In our experience, the deprotection solution can be stored at room temperature for 2–3 weeks.

5. Be very careful while using trifluoroacetic acid. Never add water into any acid; add acid to water instead.

6. The cleavage solution can be stored for up to 4–5 weeks at room temperature. It is advisable not to use cleavage solution older than 5 weeks.

7. This synthesis can be done on other scales, typically from 0.1 to 2 mmol. All reagent amounts should be scaled accordingly.

8. Kaiser test details [16]: This test is also known as the ninhydrin test. Ninhydrin, one of the components of the testing solution, reacts with the deprotected primary –NH₂ group to produce a blue/purple color. This test involves three solutions whose compositions are as follows:

   Reagent A:

   1. Dissolve 1 mg of KCN in 1 mL of distilled water.
   2. Add 9.35 mL of deionized water into the above solution to get 10 mL of 1 mM KCN solution.
   3. Remove 400 μL of above solution and dilute it with 19 mL of pyridine.
   4. Pour this solution into a small amber bottle and label it "A."

   Reagent B:

   1. Dissolve 1.2 g of ninhydrin in 20 mL of ethanol.
   2. Pour this solution into a small amber bottle and label it as "B."

   Reagent C:

   1. Dissolve 16 g of phenol in 20 mL of ethanol.
   2. Pour this solution into a small amber bottle and label it "C."

   Test procedure

   1. Add several resin beads (10–20) into a small test tube.
   2. Add one drop of each of the reagents A, B, and C using an eyedropper or plastic transfer pipette.
   3. Heat the glass tube in a sand bath set to 110 °C for 2–3 min.
   4. Observe the color change.

   Test results

   1. Blue/purple resin and solution: free amine is present; the resin is deprotected.

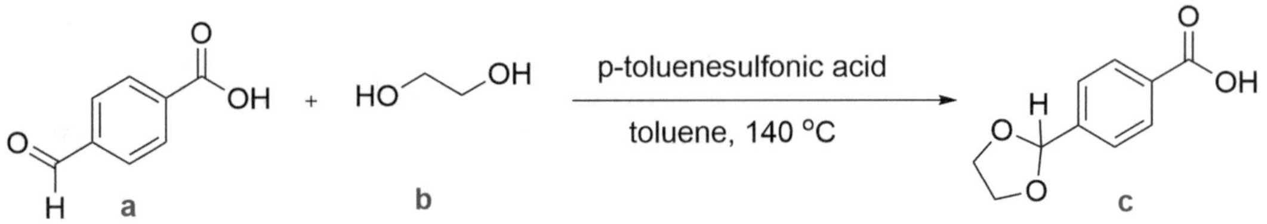

Fig. 4 Synthesis of 4-(1,3-dioxolan-2-yl)benzoic acid

Light yellow: no amines are present; the coupling is complete.

9. Peptides are drawn with the N-terminus on the left and the C-terminus on the right. Fmoc-based SPPS is carried out starting with the C-terminal amino acid and ending with the N-terminal amino acid. Thus, for peptide **D**, Fmoc-Glu(OtBu)-OH is added first.

10. The amount of DMF in the coupling cocktail used to dissolve the amino acid will vary according to the scale of synthesis. The solution should be at least 2–3 mm above the resin bed after it has been added to the peptide synthesis shaker vessel.

11. This synthesis of 4-(1,3-dioxolan-2-yl)benzoic acid (**c**) shown in Fig. 4 is a modified version of a literature procedure [17]. Add 5.0 g of 4-formylbenzoic acid (**a**) to a 100 mL round-bottom flask and dissolve it in 30 mL of toluene. Add 2.2 mL (1.5 equiv.) of ethylene glycol (**b**) and 0.28 g (0.05 eq) of *p*-toluenesulfonic acid into it. Attach a Dean-Stark trap and a condenser to the round-bottom flask, and heat the reaction mixture at reflux (set oil bath to 140 °C) for 16 h. Monitor the reaction via thin-layer chromatography (TLC), using a mobile phase of 20% ethyl acetate in hexanes visualizing using a UV lamp. Once complete, remove the round-bottom flask from the oil bath, and allow the reaction mixture to cool to room temperature. White crystals of 4-(1,3-dioxolan-2-yl)benzoic acid will precipitate from the reaction mixture. Collect the crystals by filtration and dry them under vacuum. The product (**c**) can be stored indefinitely at room temperature in a 20 mL glass vial.

12. As an alternative, 4-formylbenzoic acid can also be used without protection in the final coupling step. In this case, triisopropylsilane should not be used in the cleavage solution because it can reduce the aldehyde. In our experience, using 4-(1,3-dioxolan-2-yl)benzoic acid instead of 4-formylbenzoic acid increases the yield of peptide **C**. Note that the acetal protecting group is removed in the final cleavage step; no additional deprotection steps are needed.

Fig. 5 Synthesis of *S*-benzoylthiohydroxylamine

13. In our experience cleavage time should be kept to less than 2 h to avoid reduced yield of the product.

14. The precipitated peptide can be difficult to filter using vacuum filtration. We recommend the use of a small filter funnel with a glass frit of coarse or medium porosity. When using filter paper and a Buchner funnel, use low vacuum and filter at a slow rate to minimize loss through the sides of the filter paper.

15. The mass spectrometer employed here is an Advion ExpressIon Compact Mass Spectrometer. An ESI probe was used with negative-ion source settings. The mobile-phase solvent used was 0.1% NH_4OH in LCMS-grade water.

16. The lyophilized product is a fluffy, low-density powder and should be handled with extreme care to avoid loss. Press the powder together with spatula before transferring it.

17. To obtain dry DMSO, remove about 10 mL of DMSO from a bottle of anhydrous DMSO and add it to a 20 mL glass vial. Add activated 3 Å molecular sieves to the vial, then flush thoroughly with nitrogen gas, and seal the vial. Leave it sealed for 24 h before use.

18. The procedure below for synthesizing *S*-benzoylthiohydroxylamine (**f**) shown in Fig. 5 is adapted from a literature procedure [18]. Transfer KOH (560 mg, 10 mmol) to a 100 mL round-bottom flask and add 15 mL water to it. The flask will be warm, so allow it 10–15 min to cool down to room temperature. Next, add thiobenzoic acid (**d**, Fig. 5) (690 mg, 5 mmol) and hydroxylamine-*O*-sulfonic acid (**e**, Fig. 5) (565 mg, 5 mmol) to the above solution. Stir the solution at room temperature for 20 min. The *S*-benzoylthiohydroxylamine product will crash out as a white precipitate. Collect the solids by filtration through a coarse glass frit, and wash them with water (2 × 20 mL). Collect the crude product and dry it under vacuum at room temperature for up to 2 h. The product (**f**, Fig. 5) decomposes at room temperature, so purify it using flash chromatography (100% DCM) on silica gel shortly after making it. The product is visible by TLC using 10% ethyl acetate in DCM as the mobile phase with visualization by UV light.

19. Peptide **D** is prone to hydrolysis, which is accelerated under basic conditions. The pH should be closely monitored while adding the K_2CO_3 solution. An alternative procedure is to dissolve the lyophilized peptide directly in pH 7.4 phosphate buffer solution (0.1 M) at 10 mg/mL. Often the solution develops a faint yellow color if the pH exceeds 8.0, indicating that the hydrolysis reaction is occurring.

20. Na_2S should be massed out in a glove box and stored in a sealed vial that is opened immediately before use.

Acknowledgments

This work was supported by NSF Grant DMR-1454754 and the Virginia Tech Institute for Critical Technologies and Applied Science (JFC12-256).

References

1. Wang R (2012) Physiological implications of hydrogen sulfide. A whiff of exploration that blossomed. Physiol Rev 92:791–896

2. Wang XH et al (2013) Dysregulation of cystathionine gamma-lyase (CSE)/hydrogen sulfide pathway contributes to ox-LDL-induced inflammation in macrophage. Cell Signal 25:2255–2262

3. Szabo C, Wang R, Jeschke MG (2009) Hydrogen sulfide is an engeneous stimulator of angiogenesis. Proc Natl Acad Sci U S A 106:21972–21977

4. Zhao W, Zhang J, Lu Y (2001) The vasorelaxant effect of H_2S as a novel endogenous gaseous K_{ATP} channel opener. EMBO J 20:6008–6016

5. Bhatia M (2015) H_2S and inflammation: an overview. Handb Exp Pharmacol 230:165–180

6. Abe K, Kimura H (1996) The possible role of hydrogen sulfide as an endogenous neuro modulator. J Neurosci 16:1066–1071

7. Yang G et al (2007) H_2S, endoplasmic reticulum stress, and apoptosis of insulin-secreting beta cells. J Biol Chem 282:16567–16576

8. Hoare TR, Kohane DS (2008) Hydrogels in drug delivery: Progress and challenges. Polymer 49:1993–2007

9. Rajangam K et al (2008) Peptide amphiphile nanostructure-heparin interactions and their relationship to bioactivity. Biomaterials 29:3298–3305

10. Gao J, Zheng W, Zhang J, Guan D, Yang Z, Kong D, Zhao Q (2013) Enzyme-controllable delivery of nitric oxide from a molecular hydrogel. Chem Commun 49:9173–9175

11. Hartgerink JD, Beniash E, Stupp SI (2001) Self-assembly and mineralization of peptide-amphiphile nanofibers. Science 294:1684–1688

12. Legigan T et al (2012) The first generation of β-galactosidase-responsive prodrugs designed for the selective treatment of solid tumors in pro-drug monotherepy. Angew Chem 51:11606–11610

13. Fleming S, Ulijn RV (2014) Design of nanostructures based on aromatic peptide amphiphiles. Chem Soc Rev 43:8150–8177

14. Carter JM et al (2015) Peptide-based hydrogen sulphide-releasing gels. Chem Commun 51:13131–13134

15. Foster JC et al (2014) S-aroylthiooximes: a facile route to hydrogen sulfide releasing compounds with structure-dependent release kinetics. Org Lett 16:1558–1561

16. Wellings DA, Atherton E (1997) Standard Fmoc protocols. Method Enzymol 289:44–77

17. Pezacki JP et al (2012) Silicon and silicon oxide surface modification using thiamine-catalyzed benzoin condensations. Can J Chem 90:262–270

18. Zhao Y, Wang H, Xian M (2011) Cysteine-activated hydrogen sulfide (H_2S) donors. J Am Chem Soc 133:15–17

INDEX

Kanika Chawla (ed.), *Biomaterials for Tissue Engineering: Methods and Protocols*, Methods in Molecular Biology, vol. 1758,
https://doi.org/10.1007/978-1-4939-7741-3, © Springer Science+Business Media, LLC, part of Springer Nature 2018